FOOT PRINTS OF RED, WHITE, AND BLUE

Carl L. Adams

Library of Congress Cataloging-in-Publication Data:
Adams, Carl L., 1948 -
Foot Prints of Red, White, and Blue
ISBN 978-0-9833259-2-5

The back ground for the book cover image was adapted from image designs by Lizard and MB Photos. The book cover with title design is by the author. Some cover and interior photos were obtained from the United States National Archives and Records Administration and the Australian War Museum. All images are properties of, used by permission, and/or are licensed to, the author.

This book was self-published by the author in association with Dream Weaver Ministries, LLC, Springfield, Ohio, USA www.dreamweaverministries.com The text was updated on 12 December 2012. Proof-editing was accomplished in part by Crossroads Editing, www.crossroadsediting.com

Disclaimer: All of the stories contained within this book are true stories, gleaned from the author's personal files. Some stories contain only first names of characters; those names have been changed for protecting their privacy. Stories which may parallel and/or match identical events to that of others, living or deceased, in society anywhere are purely co-incidental.

Dedication

This book is dedicated to God Almighty, His Son, Christ Jesus, and The Holy Spirit. Without God, I would be totally lost. Without Christ, I would have no pathway of redemption. Without The Holy Spirit, I would have no peace within.

For:
All the GIs and veterans out there
who fight, and have fought, for freedom;
May God bless and protect you all!

"Blessed are the peacemakers,
for they shall be called the sons of God."
(Matthew 5:9)

For:
All my friends and family;
Who have long been requesting
this book – enjoy!

For:
Karen Haebick;
Who helped trace the
foot prints of my father during
World War II near Australia;
Thank you!

Other books by this author:

God's Dream Catcher
God's Dream Catcher Rising
*God*Prints*
Traveling Lighted Pathways
Wanted: Lost Souls
OH MY GOD!
DreamWeaver Storybook
Issues of Life Handbook

www.carl-adams.com
www.godsdreamcatcher.com
www.dreamweaverministries.com

Contents

The Author

Carl Adams is a retired U.S. Navy Logistics Specialist with over thirty-five years of service. He has also served in the U.S. Army and the U.S. Air Force. He served aboard the Aircraft Carrier USS *Independence* as a Flight Deck Troubleshooter and also flew with the 711th Special Operations Squadron as an Aerial Gunner.

Having graduated from Pensacola State College in 1982 with a degree in Liberal Arts, he has acquired many other government training courses along his military career path. Those courses further honed his knowledge, skills, and abilities in the arts of readiness, improvising, and contingency planning. Those skills propelled him to excel in the field of weapons system maintenance technology and achieve Level III Certification (highest) in the field of Military Acquisition Logistics. His achievements toward the end of his career would earn him the title of the 'Foremost Technology Transfer Specialist within the Department of Defense.' That title would be given to him by the Office of the Under Secretary of Defense for Science and Technology in 2003.

During his long career with the military, some of his direct personal services provided were Aircraft Mechanic and Electrician, Truck Driver, Infantryman, Aircraft Trades Instructor, Aircraft Examiner, Production Controller, Equipment Specialist, Aerial Gunner, Training Specialist, Logistics Specialist, and a few other specialties (not for disclosure) along the way. He has been in submarines, ships, missile silos, tanks, armored vehicles, and aircraft for deployment of military missions. He has repaired various fixed/rotary-wing military aircraft and avionics equipment, as well as having been transported in them for mission deployment. He has also been a jet backseater

(Observer) and an AC-130 Spectre Gunship Aircrewman (Aerial Gunner). On the flip side of refurbishing government weaponry, he has also demolished the weapons equipment of select non-friendly agencies in sanctioned special operations tactical missions.

In the course of his military career, he sustained a traumatic head injury, nearly losing his life, while serving in the U.S. Navy. Then while serving in the U.S. Air Force, he once became lost at sea, and all but one piece of his survival equipment malfunctioned (and then also nearly losing his life). These two close call stories with death are captured in some of his books.

His last duty station was Wright-Patterson Air Force Base (WPAFB) in Dayton, Ohio, attached with the Joint Depot Maintenance Activities Group. For eleven years, his assignment was with two joint technology service units: the Joint Depot Environmental Panel (JDEP) and the Joint Technology Exchange Group (JTEG). Through JDEP and JTEG, he interfaced with all the facets of military, industry, government laboratories, the National Aeronautics and Space Administration (NASA), and academia. Their mission was seeking improved methods for cost-effective enhancements of U.S. military weapon system sustainability.

During his assignment at WPAFB, one of his collateral assignments was interacting as a liaison with the acquisition community on common environmental issues which interfaced with the sustainment community of the Department of Defense (DOD). Those interfacing issues led to enormous cost savings for both of those DOD communities, and also for the establishment of a working group to explore the founding of a new organization. This new organization would address issues common to both acquisition and sustain communities within the DOD. Through his efforts, and suggestion for naming of the new organization, the Joint Acquisition Sustainment Pollution Prevention Activity (JASPPA) was established in the mid-1990s.

Since 1999, he has been engaged in jail and prison ministry. In those years, his involvement in ministry to the incarcerated has taken him into nine different correctional institutions, including Ohio Death Row. He has personally developed, and instructs, a life skills mentoring class for the incarcerated. He serves as an inmate mentor, a detention center chaplain, and is a certified chaplain with a major church denomination chaplain's commission. In addition to ministry to the incarcerated, he has worked with troubled youth since 2004 in a program called *H*elping *O*ur young *P*eople conn*E*ct with God (HOPE). HOPE sessions address many of the unique problems youth face today. In each session, God will always be an added resource for resolutions to those problems discussed.

Just prior to his retirement from the Navy, he encountered an out-of-body experience, which totally reset all of his life values (this experience is also captured in some of his books). In the resetting of values, the course of his life was changed toward dedicated ministry after retirement. In his later years, he has taken to writing books. He desires to impart the knowledge of what God has done for him to all who would read his stories. He believes a life legacy is worth leaving behind in books; values can never be placed on those inheritances. They will live on, becoming greater in value as the years pass, helping memories never to fade, and assisting others in finding their destinies.

Foot Prints of Red, White, and Blue is the author's ninth published book. As this book was being written and published, he began manuscripts for writing and publishing other books. He has established his own websites and has Author Webpages on Filedby.com and Amazon.com. Some of his Internet website addresses have previously been listed should anyone desire looking into them. Oh, and by the way... yes, the author claims there really are aliens at WPAFB – masses work for many of them in secret!

Prologue

There are so many books published containing stories laced with blood, guts, and foul language, especially books on war. While those true grit events are naturally a part of the military way of life, there are a great many in society desiring to connect with that life, yet be shielded from much of its abrasiveness. This book, however, was scripted with a different concept: to provide those readers with quality stories of military adventure, combat, war, and perilous zones in between – without excessive harshness, gore, and tainted language.

In the shadowy danger zones of life, especially in warfare and with military operations, there are times when an unseen *Someone* is watching over the sentinels of freedom. Those times from personal protection to blatant hell on earth can be captioned with reverence for the Almighty. This book contains a few of those stories ranging from humorous to serious, from planned to unforeseen, and from routine to the heroic.

This book contains several short stories of military life and operations, mostly from my own experiences. A few of the stories are gleaned from friends and relatives. One story is a novelette of the World War II era and is several pages in length. Other stories range from the American Civil War to the Vietnam Era. Many of the stories contain a Christian spiritual principle in them, and some do not. All of the stories, though, do include some amounts of significant life principles embedded within their pages.

There are multiple intentions for this book. One purpose is to express my appreciation for the Lord God sparing my life in the situations mentioned within some of its stories. There have been many other times, though, throughout my life when He has done the same and are not mentioned herein. Another purpose for this book is bringing attention toward the lives of others and what God has done in their lives.

The ***primary*** purpose for this book is to give honor to the Lord God Almighty for His protection over those sentinels of freedom for this great land we call The United States of America. It is He, the Lord God, who watches over the guardians of our freedom and our land – and we, His people, should ever be thankful for His guiding light.

May God Bless America!

Patriotism means to stand by one's country. It does not mean to stand by the president or any other public official.
Theodore Roosevelt
26th President of the
United States of America

There are some who've forgotten why we have a military. It's not to promote war; it's to be prepared for peace...If we lose freedom here, there is no place to escape to. This is the last stand on earth.
Ronald Wilson Reagan
40th President of the
United States of America

If we will not be governed by God, then we will be ruled by tyrants.
William Penn
Champion of democracy
and religious freedom

BOOT PRINTS

One night I dreamed of walking with the Lord along a beachhead in the aftermath of chaotic battle. Across the skies flashed scenes from many past engagements.

In each scene was noticed boot prints in the sands. Sometimes there were multitudes of *boot prints*, but always overlaid with a larger set of *foot prints*. At other times there were only two sets of imprints; one set of *boot prints*, and one set *foot prints*. Then, there were times of only one set of *foot prints* which could be seen.

These sights troubled me because I noticed that during certain times in my life, were those when I was suffering in foreign lands, and I could see only one set of *foot prints*. Or, there were times when my comrades and I would be engaged in conflicts, and I would see the multiple *boot prints* overlaid with the larger *foot prints*.

So I said to the Lord, 'You promised me, Lord, that if I pledged allegiance to my Country and You, and then followed You, that You would be with me always. But I have noticed during the most trying phases of my service and my life, there has been only one set of *foot prints*. Whose were those lone *foot prints*, and why when I needed you most, have You not been there for me, and with me?'

The Lord replied, 'The times when you have seen only one set of *foot prints*, warrior child of mine, are when I lifted you from harm's way. The times when those *foot prints* overlaid the *boot prints* are when I fought for you, and along with you and your comrades. I will never leave you, nor forsake you, for I am your Everlasting God.'

The Red, White, and Blue

When considering the colors of the American Flag, one may ponder where the colors originated. Tradition has it that President George Washington believed the stars were taken from the sky, the red stripes represented the British colors, and the white stripes signified the American colonies' secession from their home country of Great Britain.

Actually, research proves there was no official reason for the choice of colors as recorded by the Continental Congress. It is factual, though, that in 1782, Congress chose the colors of *red, white, and blue* for the Great Seal of the United States. In choosing those colors, their meanings were listed as red for valor and hardiness, white for purity and innocence, and blue for vigilance, perseverance, and justice.

Regardless for the meanings of those original colors, that 'Grand Old Flag,' continues to represent patriotism and pride in the United States of America; a land that is loved by its people, and home of the brave and the free. That image of the Grand Old Flag carries with it an imprint (or implication) of freedom to all who gaze upon its stars and stripes. Over the years, men and woman have literally wept over its sight in the far reaches of this globe. Even though those stars and stripes have been tattered and torn, people have embraced its *implied* meaning and imparted them into their hearts for solace.

In forlorn desolate outlands of captivity, ensnared peoples waited for the knights of its blood stained threads to give them hope within their souls for yet better days. They knew those knights of stars and stripes carried with them the right to be free. And, those knights of freedom would come, paying the supreme price – spilling their blood for tyranny's captives.

Regardless of anyone's religious views as to Christians settling America or not, ***history records*** that people fleeing tyranny called on God Almighty for a land of freedom, and God blessed them with America! Those who would deny this fact need to carefully research American history all over again.

'The right of the people to be secure in their persons, houses, papers, and effects, against unreasonable searches and seizures, shall not be violated, and no warrants shall issue but upon probable cause, supported by oath or affirmation, and particularly describing the place to be searched, and the persons or things to be seized' (The United States Constitution, Fourth Amendment).

This right was conceived by the forefathers of America, ***many*** who were ***Christians*** and descendants of peoples ***fleeing oppression and tyranny*** from other lands and governments. This right was set in motion by the framers of the American Constitution, who sought freedom as a way of living for ***all people*** in a ***new land*** under the blessings and ***guidance of God Almighty***.

This right was solemnly paid for with the *blood of many American GIs and patriots*, their sons and daughters then, and has continued ever since. This right has been gravely purchased for others struggling to be free in the far corners of the earth by America since it was born in red, white, and blue. Thank God *the red, white, and blue* still waves in spite of those who desire dragging it down from within and without!

May God Bless America!

Never So Few

The American Revolutionary War (which is also called the American War of Independence, and the Revolutionary War), began as a war between the Kingdom of Great Britain and the Thirteen British Colonies in North America. It ended in a global war between several great European powers. The Thirteen British Colonies in North America (the American

Colonies) formally declared their independence from Great Britain on 4 July 1776. Yet, the first shots of freedom were actually fired many months before on 19 April 1775. The conflict for *sovereign* independence lasted until 3 September 1783.

The total casualty loss resulting from the American Revolutionary War is unknown because no reliable statistics are in existence for those numbers. As was characteristic with the wars of that era and before, disease typically claimed more lives than battle. American statistics (rough at best) reveal the estimated total American military casualty figure was as high as 50,000 Revolutionaries. Of that number, about 25,000 died during active military service. Of those deaths about 8,000 were killed in action, while the remaining 17,000 deaths were from disease. Historian Joseph Ellis claims that General George Washington's decision for having his troops inoculated against smallpox was probably one of his most important decisions made during that war.

The American Revolutionary War became very expensive for every family and every nation involved. Its costly scars left indelible imprints on nations around the world. Yet, its conflict would birth freedom for a brave new world, the United States of America – *the land of the free and home of the brave*. Little did those brave pilgrims know, their descendants would become known the world over as the USA; *the red, white, and blue*; a people who import freedom and justice to other lands.

World War I was predominantly called the Great War, or the World War, from its occurrence until 1939. After that year, it was called the First World War or World War I (WWI). Its major conflicts centered in Europe, with the first conflict beginning on 28 July 1914. The war officially ended on 11 November 1918.

The total number of military and civilian casualties in WWI amassed over 37 million. With over 8 million killed in action (KIA), over 21 million wounded in action (WIA), and over 7 million missing in action (MIA) – its conflicts ranked among the deadliest in human history. About two-thirds of military deaths in WWI were directly battle related. Diseases, including the flu, caused about one third of total military deaths for all combatants. WWI was very costly for every family and every nation involved. While assisting the great European powers in their conflicts, American expeditionary doughboys left their patriotic footprints of freedom from their fighting excursions all over European theaters of conflict.

The roots of conflict for World War Two (WWII) can actually be traced to 7 July 1937, with the Second Sino-Japanese War; a war between Japan and China. Although Japan was already at war with China, WWII is generally said to have begun on 1 September 1939 with the invasion of Poland by Germany. After the assault, declarations of war on Germany were issued by France and most of the British Empire. After years of appeasement toward Adolf Hitler's insistent goals of European domination, the undeniable path to the Second World War was ignited with those first shots fired when Germany entered Poland in 1939.

Even after Europe became aflame with war, America stayed out of the conflict until the Japanese attacked Pearl Harbor on 7 December 1941. After America entered WWII, fighting and espionage took place in nearly every continent and ocean on planet Earth. WWII officially ended on 2 September 1945 when the Empire of Japan signed its surrender on the deck of the USS *Missouri* in Tokyo Bay.

With WWII impacting nearly every continent and ocean on planet Earth, its conflicts claimed record staggering figures for KIA, WIA, and MIA. WWII casualty statistics vary; estimates of total dead (military and civilian worldwide) range from 62 million to over 78 million. The total military KIA ranges from 22 to 25 million. WWII has been declared the *largest and most deadly war waged in all of human history*. It has been estimated that *three percent* of the world's population, during that time frame, died as a direct result of the war.

Beyond WWII, numerous other wars could be mentioned with their statistics. Especially for the USA, casualty statistics for the Korean War, Vietnam, and the War on Terror could be mentioned. However, for impact and keeping the dialog relatively brief, only the rounded-off casualty statistics for the major three wars of the USA up to the date of publishing this book have been mentioned. Beyond the American Revolutionary War, the USA contributed (in a figurative sense), the hearts and souls by the hundreds of thousands of its sons and daughters to fight despots and tyrants all over the globe.

According to Wikipedia, at the time of drafting the manuscript for this book, there are some surprising casualty statistics for the USA military. In all of the wars and conflicts the USA has become involved with since 1775, the USA has sustained approximately 2,489,335 military casualties. This number includes 848,163 KIA, 1,529,230 WIA, 38,159 MIA, and 437,421 other (mostly disease). The grand total comes to 2,852,973. In 237 years of war and conflicts around the globe, the USA has sacrificed 2,852,973 of its *military* sons and daughters in combat. Those sacrifices were for securing the freedom of their own homeland *and also* securing the freedoms of other lands.

Most everywhere the American doughboys and GIs have stepped, they were welcomed. Why? The oppressed were very well aware those footprints carried the stamp of freedom! The oppressed recognized the imprint of *the red, white, and blue* – even though it came shrouded in olive drab, desert tan, navy blue, or ops black. The oppressed were also very well aware that freedom was too often paid in full with the blood from an American doughboy or GI. Thus, they were glad to render assistance for furthering the cause of freedom.

During the largest and most deadly war waged in all of human history (WWII), for the USA, that armed conflict lasted 1,365 days with 65 major battles and engagements on two fronts of the world's globe. Approximately 16 million Americans served in the armed forces during WWII. With approximately 416,800 listed as KIA and MIA, and over 700,000 WIA, the casualties totaled *over one million*. Those grave statistics averaged 305 casualties every day during the war for the USA. Thus, from those statistics came a creative concept for the military poster image on the beginning page for this entry. That world famous image was taken by Associated Press War Photographer Joseph Rosenthal. The military poster is titled:

'Iwo Jima 1945
Never Have So Many
Owed So Much To So Few.'

When considering the granting of freedom from tyranny, there are only two rare breeds of individuals who have given their lives for it: the *soldier* and the *martyr*. Both willingly give for the cause they believe in. When contemplating the equation of freedom, it will always be found that *Freedom is not free*. Someone somewhere all through the course of history willingly gave their lives for freedom from the despots. Comparatively, the U.S. soldier died for freedom all around the globe because they believe in a world free of tyrants.

On a spiritual level, Christ, the Son of God, died for the sins of the entire world, because He loved it; His mission and purpose was restoring those who would receive redemption, and be eternally free.

'All that is necessary for the triumph of evil
is that good men do nothing'
(Edmund Burke)

Ghost Stories

This story is actually an introduction to the two short stories which follow after it. Those two short stories are my personal testimony of how God spared me from certain death while in the military. There were other times, but those two times were very harrowing incidents. The events of those days are embedded in the memory banks of my mind. This entry of *Ghost Stories* is a prelude to the first near death event. It didn't have to be included in this book, but its dialog was relative for it shows God does keep a watchful eye over His people.

These three stories of *Ghost Stories, Crushed,* and *Lost at Sea,* are in some of my other books. Those who have read them may consider these particular stories redundant. However, I am most thankful to God for what He has done for me, especially in sparing me from death. Therefore, since not everyone will read all of my books, it is my desire to place my personal testimonies in as many of my books where they can be applicable in the book themes or story lines.

Most of us are intrigued by ghost stories, whether we believe in ghosts or not. The same goes with evil spirits, regardless if we believe in them or not, we are still curious about them. Associated with ghosts, is usually always death; and death has most commonly been personified as some figure or fictional character from mythology or culture since the earliest days of storytelling. Because the reality of death has had significant impact on the human psyche, the characterization of death as a very vivid entity is a concept which has existed in many societies since the beginning of recorded history.

Variations of skull faces and grim reapers, for centuries, have been popular items of interest among the living because they personify death itself. Images and shadowy sketches of skull faces and grim reapers have reportedly been seen by many people as premonitions prior to seeing death, or facing their own death.

The skull face has even been used as an emblem on military uniforms. The most popular example was the Totenkopt used on the German military uniforms of World War II. In more modern times, the combination of a skull face and a grim reaper was used for a unit emblem for one of the military units I served time with, the U.S. Air Force 711th Special Operations Squadron. As an aircrew member of a 'Spectre' Gunship Unit, we had mottos of, 'You can run, but you can't hide,' 'You can run, but you'll only die tired,' and 'Anywhere - Anytime.'

And now, for my own personal ***ghost*** story...

Waking from sleep one summer Friday night in 1968, there appeared a grayish object in the U.S. Navy barracks where I slept. The rooms of this barracks at the time bunked four sailors to a room. This particular night the other three roommates were not there. I had bunked sometime around 2230 hours (10:30PM), and a vision came to me later. This vision occurred sometime before midnight, for I'd questioned my roommates the next day about what time they came in. One came in just after midnight and the others came in about 0100 (1AM). I remember when seeing this object, the other things in the room could be seen, and those roommates were not there.

This object appeared near the head of my bunk. When it did, grayish outlines of its face could be seen in the dimly lit room. Focusing my eyes, its appearance was that of a skull in shades of gray covered with sparse amounts of thin skin stretched over only parts of it. The skull had no eyes, just empty eye sockets. Its lower jaw, with broken teeth, was slowly moving up and down as it slowly floated closer to me. So frightened at the sight of this, my ability for movement was frozen.

This horrendous looking thing came so close to my face, its breath was felt coming from its mouth with its opening and closing. As its warm breath was felt flowing over my face, its putrid odor penetrated my nostrils. Then the skull began hovering over and around my face even more closely; and I still could not move. Paralyzed by the sight and stench, the sensation of evil could be felt emanating from this skull. A feeling began instilling within me, that if this thing were to bite me, evil would possibly either enter me and/or torment me.

Needless to say, praying to God immediately began. Pleading in the name of Christ Jesus for God's protection from whatever this evil thing was, I sought God for making it go away. Then, everything went completely dark, and the next thing known was that morning had come. Sitting up on the edge of my bunk, I thanked God for Him taking the thing away.

This horrid skull face that night had appeared twice before during this same year; the first time it appeared at the barracks room window. The second time it appeared at the window and slowly moved between the window and my bunk. Both times its mouth was seen moving, and both times I tried disregarding them. With this third appearance, however, it was very personally directed toward me in coming right up to my face and nearly touching me.

As previously mentioned, skulls and crossbones are usually associated with, and are also usually symbols of, death or dying. It is my opinion the skull was symbolic of death actually knocking at the door of my life. Why do I state this? Only a few months into the future, my head would become nearly crushed in an aircraft maintenance incident.

Later in the fall of that same year, while still serving in the U.S. Navy, I was awakened from sleep one night to the appearance of a black object in my barracks room. Everything was much the same as before, except this was a Saturday night and I had bunked about 0100. One roommate was on Barracks Watch Duty that night, another one came in around 0400, and the other one was away for the weekend. While seeing this object, other furnishings in the room could also be seen, and those roommates were not there.

This black object came through the locked room door, positioned itself at the foot of my bunk, and turned facing directly at me. Its appearance was that of a hooded cape like that of the Grim Reaper in shades of black and gray, with no face in the hollow hood. The swaying cape began a slow menacing rise up over my bunk. The emanating evil spilling from its folds possessed the same strength as the skull face, but not the same intensity because of its proximity. As with the skull face, a sensation penetrated me that, if this thing came into contact with me, evil would possibly either come into me and/or torment me.

Wasting no time, my praying for God to take this thing away began. Crying out to God, I rebuked this thing in the name of Christ Jesus, and pleaded His blood over me. Then as before with the skull face, everything went completely dark. A passage of time was sensed and I awakened. Flipping on the room lights and looking at a roommate's clock, it was about 0300. One of my roommate's lockers was unlocked, and looking in, nothing was seen but his uniforms and civilian clothes. There was no other person or thing in the room which could be seen. Standing in the middle of the barracks room, thanks went up to God for Him taking that thing away.

It is my opinion evil was attempting to get at me. Thinking about the symbolism of skulls and crossbones, there is symbolism with dark capes too. Dark capes are usually associated with death or dying also, such as with the image depicted by 'The Grim Reaper.' Again, it is my opinion this dark cape, along with the skull, were symbolisms of death actually knocking at the door of my life. In only two or three months from this time, my head would become nearly crushed in an aircraft maintenance incident.

Since that time, I have seen those sightings during crucial periods of my life, especially during two types of occasions. One was during times when, after their appearance, the possibility of death had come near. In these instances, their appearances are simply an evil avenue for the tool of frightfulness to unsettle spiritual peace about possible coming death. The other occasions were when accomplishing ministry efforts for the Lord God, and the work had positive impact on others for His kingdom. In those instances, their appearances were used by evil in discouraging believers in the Lord God. God does not bring these horrid things to the sight of the believer; however to the believer who can withstand the sight of them, the Lord God permits the 'seeing' of those things as a vision. Those sightings are for the admonition of those believers who are soon to experience some evil influence coming against their lives.

Ephesians 6:12 says, *"For we do not wrestle against flesh and blood, but against principalities, against powers, against the rulers of the darkness of this age, against spiritual hosts of wickedness in the heavenly places."* With the knowledge that evil spirits can, and do, come against us, this true-life ***ghost story*** is an excellent lead-in for the next true-life story, for these events were premonitions of a coming Oh My God event – these were warnings of its impending arrival...

Crushed!

When as a young twenty year-old sailor in the U.S. Navy, I was once involved in a very strange and serious aircraft maintenance incident. Four times a few months before the event, a vision of death came to me. Three were as a skull similar as the 'Skull and Crossbones' and in one of those attempts it seemed intent on devouring me. The last vision was one like the 'Grim Reaper' and its dark cape came very close to completely enveloping me.

The ancient mariners have a legend that when seeing a ghost a sea, a disaster is forthcoming, and death may be accompanying it. Therefore, with seeing these things in a short span of just a few months, I was somewhat concerned as to what held my destiny.

During the night of 16-17 January 1969 my mother, Pauline Adams, had a very disturbing dream. She dreamed of seeing a photograph of me lying on the ground during a very heavy rainstorm. She then observed a large body of blackened floodwaters rapidly approaching the picture. Somehow within her, she knew that when the blackened floodwaters reached my picture, I would die. She suddenly awoke and, sensing imminent danger, immediately began praying and interceding to God for me.

Later that morning around 0900, one of my friends (Janice Trawick) was driving home from college. A few blocks from her home, suddenly she saw my face flash about six feet in front of her, above the hood of her car. She stated the face was as real looking as if it were actually me being there right in front of her. Janice also sensed something very wrong was about to happen to me. She also began praying and interceding to God for me.

Later that afternoon of 17 January, my mother received a phone call from the United States Navy. She was officially informed I had been hospitalized at the Pensacola Naval Hospital for a traumatic head injury sustained from an aircraft maintenance incident. The Navy spokesman informed her that my condition was stable and I was expected to recover, and that more information would follow later.

On that morning of 17 January 1969, I was a U.S. Navy Aircraft Mechanic assigned to the Naval Air Station in Pensacola, Florida. That crisp winter morning my shop was performing safety inspections on ejection seats in all squadron aircraft. All squadron aircraft were grounded as a result of a Safety Bulletin issued from the Naval Air Systems Command. At about 0830 hours I was standing on the left side of an aircraft fuselage with my feet in the aircraft 'kick-steps.' My torso was across the cockpit rail and my head was in the aft cockpit, down near the left console, between it and the ejection seat. I was performing the required maintenance as mandated by the Naval Aviation Safety Bulletin.

On completion of the task, my left arm withdrew from out of the cockpit and onto the cockpit rail for steadying myself before climbing down onto the flight line. As I did so, the left shirtsleeve of my uniform caught the canopy actuation switch and immediately the canopy began lowering. At the position of exiting, my head was instantly pinned across both temple areas to the cockpit rail by the moving canopy frame. Immediately I flung my left arm clear of the canopy switch. The canopy did not stop! Frantically searching for the canopy switch, I hoped for repositioning it to the up position. When finding it, there were no results, for the canopy continued compressing my head.

The shocking realization struck me like an explosion that, within a few seconds my skull would be crushed! I was going to helplessly die in an excruciatingly painful death!

My thoughts turned toward God as I began hearing 'splintering' and 'crunching' noises inside my head. 'Oh God,' I thought, 'I am about to be one of the aircraft mishap statistics that we heard about in Aircraft Mechanic School.' Then I screamed as loudly as possible. Maybe someone was nearby on the flight line and could override the canopy position with the external canopy switch. As the canopy continued compressing my skull, weakness and blurred vision became a part of me. Just as thoughts of passing out came, I screamed, 'Oh, God please help me! I'm dying!' Instantly, and I do mean instantly, that canopy halted, and there was no more movement! 'Oh Praise God' and 'Thank you, Jesus' were in my thoughts!

I did not dare touch the canopy switch this time. It didn't work like it was supposed to when it was really needed, and for right now, I was alive! Maybe someone would come by and find me and use the external canopy switch. Swishing and ringing noises were heard in my head. Things were getting dim, and I knew fainting was coming for me within the next few seconds. Next thing I knew, the canopy was going up, and I was falling onto the flight line ramp. In falling someone caught me while saying, 'I've got you Carl!'

Looking up into his face, it was my workshop buddy, Ron Baum. With me being very weak, he threw me over his shoulders. While Ron was carrying me toward the hanger, he kept saying, 'It'll be ok, we'll get you to Sick Bay. Just take it easy.' In a semi-conscious state, I ran my right hand across the temple areas in both sides of my head. Hearing the splintering and crunching sounds, and feeling the most intense pain of my entire life, I knew some serious damage was done. My fingers discovered large indentations deep enough to hide them in.

Once in Sick Bay, states of twilight faded in and out as some Navy Corpsmen readied me for an ambulance. It rushed me to the Navy Hospital for admittance. During the transportation time I became very weak, dizzy, sick on my stomach, and my head felt as if it were going to explode. Ringing noises were growing louder as swishing noises changed into the sound of a rushing loud waterfall. Whatever was spoken to me could hardly be heard.

Several x-rays were taken of my head and then I was placed in a hospital ward. Sleep immediately overtook me, but I was awakened every two hours for preventing coma. The next morning I managed to look into a latrine mirror and in doing so, burst out crying for the reflection that was seen. My head was very swollen, and my ears appeared to be on raised lumps on the sides of my head. The top of my head was somewhat pointed, like a cone head. 'Oh dear God,' I thought, 'If I make it, am I always going to look like this?'

That evening, a Navy Flight Surgeon discussed my cranial condition and x-ray results. The Navy Doctor informed me that my skull had sustained all the external pressure it could possibly withstand. The x-rays revealed maximum shifting, without serious damage, of the skull plates. All ligaments holding the skull plates together, were 'extruded' the maximum possible amount outward from the skull. The Navy Doctor informed me of being extremely fortunate to be alive, and it was not normally expected for someone with this type of head injury to survive. He further stated had the aircraft canopy closed just a fraction of an inch more, my head would have burst open like a crushed eggshell. If not death, then resulting brain damage would have been most certain. I did miraculously recover, but for several weeks my head was very sore and severe headaches were encountered. Medication reduced the severe headache pains as the swelling receded. Damage to my left ear, though, never completely disappeared.

The year 1972 found me working as a Navy Aircraft Electrician at the Naval Aviation Depot there at the Naval Air Station in Pensacola. One of the electrical shops, to which I was assigned, overhauled the canopy actuator installed in the aircraft which I'd worked on that dreadful day in 1969. Working with the overhauling technician, I inquired if the actuator had any internal safety limit switches for disengaging the drive motor from the actuator mechanism. Subsequent discoveries revealed the motor was capable of applying tremendous torque of several thousand pounds per square inch, and there were no limit switches designed internal to the actuator, nor within the motor. Without the external limit switches, which were located on the aircraft, the actuator motor would run until the applied torque could literally burn out the windings of the motor. In other words, that actuator motor could have crushed my head that day, but did not. Only God knows why.

In 1984 another U.S. Navy man became involved in an aircraft incident very similar to my incident. This time, however, the incident was fatal. In the same squadron I had been attached to, a sailor the very same rank and occupational trade as I had been, was working on the very same type of aircraft. He had been working in the very same position as I had been, standing on the fuselage kick-steps leaning over the cockpit rail into the aft cockpit. At the time, and according to a U.S. Navy spokesman, 'The sailor was crushed to death when the canopy of the aircraft unexpectedly became activated, thus compressing the twenty-two year-old sailor between the cockpit frame of the aircraft cockpit rail and the canopy frame. The cause of death was listed as cardio-pulmonary arrest, secondary to trauma.'

Some personal sources at the time related to me that his torso accidentally actuated the canopy switch to the closed position. Then trying to exit the aircraft (just as I had done), the lowering canopy frame pinned his chest against the aircraft cockpit rail. The sailor died from congestive heart failure. Simply, his chest was crushed to the point his heart was squeezed until it could not beat anymore. Long ago I, too, had experienced being squeezed and crushed by those tremendous pressures, and my thoughts of agonizing sympathy went out to this fallen fellow U.S. Navy man.

Headaches still plague me after all these years. In the mid 1980s' the Veterans Administration (VA) ordered some headache evaluations from contract physicians. Even though those reports supported the fact that my headaches were very possibly resulting from this traumatic head injury – all VA physicians (except one at the time of writing this book) have for years, not considered those reports for headache and head injury connection relevance. As a former U.S. Navy Examiner, I find those actions, indeed, most interesting when they have select copies of the original hospital records.

I do not like thinking about the day of my head injury. That day my head could have been crushed and I would have died. When I do think of that day, I sometimes flinch violently, and sometimes even scream. Those flashbacks take me back to experience that incident all over again. At times I have even caught myself clenching my fists until my fingers ache and begin turning colorless. Why do my fingers still change color, and flinching and screaming at times still occur? In my mind there are still vividly imprinted sights and sounds of the canopy coming down. Still can be felt, the 'crunching' of my skull. Still can be heard, the splintering noises and the waterfall sounds inside my head. Still can be seen, everything going dim; as my last thoughts were of me about to die. Still can be heard, my screaming out to God for help. I can still experience this incident all over again as if it just happened. After all those years, I can still vividly recall everything of what was thought to be my last few seconds on this earth.

In thinking about all the events of that head injury incident, there have been many thoughts if satan had actually 'set things up' for the accident that day. I can remember things back to the age of two years old. My recall ability is very vivid with details from many years past. I can vividly recall things said to me from childhood, from high school teachers, from technical schools, and from many places in all of my life. There are times, when recalling written text that I can actually 'see' the page from a book the text is located on.

Consider then, if one has such recall ability, and can recollect 'seeing' Scriptures in the mind having read in the Bible, that one could be of use to God in ministry. Consider also then, if that one with a good memory has a strong desire to serve God, that individual is most likely on 'Hell's Most Wanted' list to be permanently taken out at some appointed time by evil minions.

Thinking back over these things, the visions, and the dreams – God was in them all. God gave me visions of warning, for me to pray. God gave my mother a dream of life threatening danger coming toward me, for her to pray on my behalf. God also gave my friend, Janice, a vision of me, for her to pray for me at the very time I needed intervening help for a life-threatening situation. To me, all this has illustrated the care and sustaining grace of God, His love and concern for me, His child, and the power of intercessory prayer. As I write these things down, again I thank the Lord my God for His hand of protection and mercy. Also, I thank Him for a praying mother and for praying friends.

Thinking back over these things in later years of life, the visions and dreams leading up to this incident were impressions sent personally from God, as warnings. God sent them to my mother, my friend Janice, and myself for a purpose; that we personally pray for a critical soon coming time. God developed those imprints in my mother's dreams, and instantly right in front of Janice's eyes and mine. When experiencing flashbacks of those events from years ago, even though they are excruciating at times, they become yet more impressions, humbly reminding me of how great the Lord God Almighty truly is.

Lost at Sea

There was a time in the 1980s during my U.S. Air Force days as an Aerial Gunner that I became lost at sea. It was an aircrew sea survival training exercise gone terribly wrong; and I came close to possibly forever lost at sea (and maybe even being eaten alive). After our last parasailing exercise, we survival trainees were to be adrift alone in the ocean for the remainder of the day. After ditching into the sea, my left leg became entangled in some parachute shroud. An undertow current then threatened pulling me under during attempts at cutting myself loose from the shroud cords.

Once free, I noticed only the left side of my life vest stayed inflated. The right side, or cell, would inflate, but it was punctured and could not hold air for long. For the remainder of the day, it would require periodic inflation. Attempts for inflating the one-man life raft revealed it contained a large hole. Immediately, the life raft repair kit was accessed, only to discover it was so corroded that it was totally inoperative. Then something began striking at my feet. 'Oh great,' I thought, 'barracuda or sharks and I have to float out here with no life raft. This ain't a good position to be in.'

Next I retrieved the signal flare kit to signal for help. Opening the flare pouch revealed an unwelcome sight: the day-end of all three flares had been expended. Withdrawing them and flipping them over, produced another dreadful surprise: the night-end of the signal flares had also been expended! Talk about grave disappointment, it had arrived! Pulling out the survival signal mirror for hopefully flashing the spotter-pickup helicopter brought another great disappointment. Over seventy-five percent of the signal mirror's silver finish was missing! Still, though, I attempted its use; but all my attempts resulted in failure.

The only thing left for me to do was nothing but just float there for a while and try assessing my situation. All of us trainees were about twelve to fifteen miles out to sea, with no land in sight. Pondering over my situation, I made a shocking discovery – a surface current was carrying me swiftly away from all the other trainees. Immediately deploying the survival sea anchor, I hoped it would slow any further drifting out to sea. It worked partially. Even though a slight drift remained, at least the drifting was slower than before. This was because the sea anchor line was only about half as long as it should have been.

There was still something striking at my feet. The water was fairly calm and smooth, permitting sight a few feet beneath its surface. Whatever it was, it was not small, and appeared about my size. It was always below the water's surface, so a fin was never on the surface. Knowing that barracuda as well as sharks can grow quite large, I became very concerned. Talk about a time to pray; I prayed. I also carried a small New Testament Bible in one of my flight suit pockets. It was wrapped in Saran Wrap and in a plastic bag. At times I would place my hand over it and cry out to God for help. Over the course of events that day, whatever was striking my feet became less frequent.

Drifting for a couple of hours more, still being carried further out to sea, no longer could I see anyone else in the sea. The spotter-pickup helicopter could just barely be seen on the horizon. A storm was also seen approaching toward my direction from the southeast and the seas were beginning to become slightly choppy. All this signaled a personal 'max sweat - condition red' situation. There remained one last hope with survival equipment – a signal whistle. Sound waves from the whistle can carry over the surface of the water. With the right conditions, whistle sounds have reportedly been heard for miles. Whoever heard it would point in the direction of the sound and relay with their whistle to the next person. Eventually help should arrive (in theory during training scenarios anyway).

Pulling the whistle from its vest pocket, I blew hard into it. No noise, and nothing but my own hot air came out of it! Totally disgusted looking into it, my panicking eyes discovered the internal ball was missing! Not believing all this was happening to me, I floated helplessly in the ocean for the next several hours on one half a life vest part of the time, and with NO OTHER survival provisions. Something still occasionally struck my feet, so at times I would draw them up and hold them with my hands into the fetal position. This reduced my size profile under the water, and was difficult to float in this position with only one-half of a life vest. Whatever was striking at my feet eventually ceased striking me. It could have been a dolphin, but its shape was not visible. My fears, though, were that this was a predator, and it was just playing with its next meal, just as cats will do.

For the next several hours, nothing could be seen but sea and sky – no one, not anything. Prayer was on my lips, though, that they (the survival team) would find me somehow. These were days before Global Positioning System (GPS) equipment was authorized in general for individual troops. Cell phones had just come out as large 'car phones' so they were somewhat rare items, and not authorized for use during training. Without GPS equipment, without a cell phone, a storm on the horizon, nothing in sight, no survival equipment (and the last piece of it was only half there), I had to look at things logically – I was simply just *'dead in the water.'*

The maritime facts are that if one is lost at sea, and is not found within the first twenty-four hours, the statistical chances for recovery are radically reduced. I knew that surviving the night was a possibly. However, come morning would be a different scenario. If whatever was striking my feet were a predator, it was playing with me, and I was going to be its next breakfast. Out there in the open water of the vast sea, those thoughts were not comforting at all. Again, as logic would dictate with my prevailing circumstances, I was simply just 'dead in the water.'

Stop here, and seriously consider something. Recall any seascape seen in picture, photo, or even a movie; one in which seascape contained nothing but the ocean and sky for as far the eyes could see. That image would possibly very closely resemble the scenery totally surrounding me that day. Swirling 360 degrees in the sea would be the view similar that was coming to my eyes. With a storm coming in, a predator for an occasional unwanted visitor, and virtually no survival equipment – my situation was certainly hopeless.

However, God is the Master of hopeless situations. Even when all else fails, He never fails!

Eventually, the spotter-pickup helicopter flew by. It flew diagonally toward me, not at me. Therefore, I knew it had not spotted me in the lightly choppy seas. Splashing and waving my arms as much, as hard, and as high as possible, the helo came rolling around; its circling signified that I was spotted. Within a few minutes the safety lifeboat appeared alongside and the helo flew off. Once inside the watercraft, the crew threw blankets around me, for the evening hours had grown cold. The feel of something solid underneath me, and warmth covering me, was a most welcome sensation from the desolation of the sea.

I had been the last one to be picked up. There was only one other aviator in the safety lifeboat. Sitting right behind the boat pilot and his first mate (crewman), their conversation came into my hearing. One of them mentioned 'they lost this guy and have been looking for him for a while.' When aware of my listening, those crewmen totally dropped that avenue of discussion.

The next morning a senior school official apologized for this 'mishap' stating it was statistically not supposed to happen. He stated steps were being placed in motion so something like this would never happen again. With all the quality assurance system of checks and balances, this bizarre training 'mishap' was never supposed to happen, but it did; and it happened to me. The Lord God was with me that day, and He saw me through the situation. I was lost, but He caused me to be found!

That day of this aircrew water survival training mishap taught me a very valuable lesson. When flying, I always carried some of my own personal survival equipment along with me!

In all the years of thinking about how those events happened that day, one thought never crossed my mind until about twenty-five years later when talking with a Prison Chaplain. One of the Prison Chaplains where I volunteer came up with a unique concept. Chaplain Arthur Baron came up with the possibility that the crewmen responsible for loading the survival pack may have conceived a plan to place most of the defective equipment in that one pack (which became mine). Then inform the rescue spotters to watch for the one trainee splashing most in the water. However, someone else may have had somewhat of a sadistic thought process, and placed all the remaining defective equipment in that one very same survival pack. The results being the rescue spotters would see someone ***desperately*** troubled in the water.

What was probably not taken into consideration was the possibility of strong surface currents – one which rapidly separated me from everyone, and carried me far out to sea. In reality, satan could have orchestrated the survival pack events one day, and then the sea current events the next day. Thus, those events placed me directly in harm's way for becoming totally lost at sea with no survival equipment. Those events could have been orchestrated by evil, fostered by satan, in attempts for permanently taking me out of this life again; and far out from sight of everyone this time!

There are some analogies with this true-life story, with other things in life, and spiritual things as well. As we travel through this journey of life, sometimes we ride high above seemingly everything. Life is good, the view is great, and the ride is smooth sailing. We may have attained this position by efforts all our own, or with the help of others. The danger comes with the temptation of riding high and looking down on everyone else in life.

Sometimes, though, we arrived in those high places of life with offers from the evil one of this world. After a while of high-rise living, disconnects can occur, and we come crashing down. Within the impact of confusion, we can become dangerously entangled with the very means which kept us aloft. Then will come a time for 'cutting the cords' or the undertows for the cost of living high will drag us into a drowning death. Those situations are very simply, a cut-loose-or-go-under situations; sink-or-swim, live-or-die circumstances.

Then can come tumultuous times when all the things we have acquired for our survival, can utterly and completely fail us. The life rafts of stashed savings, self-help programs, and whatever we engineer to save us in disaster, may not keep us afloat. The security within them may not hold. Even the various life raft patch kits we have for emergencies can corrupt with time, and be rendered of non-effect when needed.

As we try lighting off flares of help for signaling others, those, too, may not have been carefully planned for. The signals we use for attracting attention may actually be void of any serviceable content. Or, those signaling tools may have been given to us by others who were jealous of us, and therefore, may not even be applicable in our situations – thus our predicaments will go unnoticed.

We may actually be flashing the mirror of self-image that no one cares to see anymore (for we have actually worn it out). The sea anchor of talking things out and partaking in formalized structured programs can be of some assistance. However, if the sea of life has a strong swift current, it can still rapidly sweep us away from every program, and every one close to us. The whistles of survival, acting as screams of last efforts for help, can sometimes be nothing but warm to hot air spewing forth into a vast sea of nothingness – falling on deaf ears. We then, can find ourselves left all alone with nothing but defective pieces of survival, as a storm gathers on the sunset of last hopes for living.

It is then, in the barrenness of absolute despair, the Lord God hears our cries. He alone has ears to hear the cries of His people. It is He who can see us when no one else can. It is He who can, and will, come along side for comfort. It is He who can, and will, pick us out of the raging sea of desolation. It is He, the Lord God Almighty, who can, and will, save us from ourselves. It is He who can restore what we have broken. It is during those desolate times when we think there is no way out of our Oh My God situations that we need to be thinking outside of our boxed-in circumstances. We need to look up and out of those boxes into another dimension – God's dimension. It is there, in that Heavenly Dimension of God, where His Divine Assistance is always found.

God spoke to the Prophet Isaiah saying, *"When you pass through the waters, I will be with you; and through the rivers, they shall not overflow you. When you walk through the fire, you shall not be burned, nor shall the flame scorch you"* (Isaiah 43:2). It is within God that we find the help we need when all else fails us. Jesus said in Matthew 28:20, *"...and behold* [take note; listen], *I am with you all days, even to the completion of the eon"* [longest division of geologic age or world] (Exegeses Bible). The Lord God is on our side, and will be with us, even to the end. That end may be the end of the known world, or the end of our individual world (life).

That day of this aircrew water survival training mishap was actually an impression sent from God. It was given personally for me. With all the turmoil prevalent in my personal life at the time, this terrible event could have been a miserable capstone for my life. Yet, it was not. God used that 'mishap' event in my life for showing me just how small and helpless I was. God also turned that 'mishap' event around and showed me just how GREAT He is. Along with His rescue operations for me came the impression of His rescue and restoration efforts for all of fallen mankind.

This lost at sea story, and its associated personal impressions given for me, has been related to many others down through the years since it happened. In relating this story, its message has gone on to become further impressions to many wandering souls lost in the sea of life. Its message has brought some lost souls into finding God, and gathered some wandering souls back to Him.

Lost at sea, I may have been in a hopeless situation – but God changed things. God caused me to be found and to one day begin frequently telling this story. Then, in relating it, its words have reached into the hearts of people needing to hear about true hope in this life, and helping them find Redemption. Souls may be totally lost on the sea of life, but they can be found and completely restored through His Rescue and Salvage Operations!

The Trash Pile

My grandfather, Richard Adams, never talked much about his life, so what little I did hear came from my father. When Grandpa was a young man, he joined the U.S. Army during WWI. Sometime during his enlistment he came down with pneumonia so bad he was hospitalized. His condition worsened until the doctors thought he actually died from it. The medics carried him outside the hospital where a pile of corpses were, and placed him in a stack of dead soldiers. Apparently a severe strain of flu had made its rounds, several soldiers had died as a result, and Grandpa was thought to be one of them.

Sometime later, a soldier on guard duty was making his duty rounds during the night. Passing nearby were Grandpa was buried in the stack of bodies, he heard some moaning coming from them. He called for help from the medics, found Grandpa still alive, and they extracted him. Grandpa was re-admitted to the hospital, administered major lung surgery, and was nursed back to survivable health. He was later given an Army Medical Discharge.

For as long as I can remember, Grandpa was frequently short of breath a lot. He could frequently be heard wheezing and his lungs would rattle when breathing. Sometimes he encountered asthma attacks and would nearly faint before getting preventative medication. After taking the medication, he would have to sit down and take it easy for a long time before he could get up again. Even though Grandpa endured much sickness in his life during WWI and after his discharge, he lived to be sixty-eight years old.

We may get knocked flat on our backs in this life, and we may encounter many sharp bends along its pathways. Circumstances may even cast us into the trash heap of those considered as discarded expendables. However, there are a few diligent people passing by who are made with the right stuff of life. They come from a different breed of humanity and they listen intently to the cries of others. They are those who make positive changes in the world around them. Their existence will spawn more life within the lives of everyone around them.

Those are the people who reach out and try saving a lost and dying humanity from ruin. They are those who try providing castaways with marks of dignity. To rise from the trash pile, into the ranks of esteemed – is a noble achievement. In this life, the Lord God wishes such for all peoples. *"...I have come that they may have life, and that they may have it more abundantly"* (Jesus in John 10:10).

We Saved the Day

Have you ever heard the cliché 'We saved the day?' Or possibly, 'They saved the day?' Usually, the phrase comes from someone, or a group of people, having accomplished something vitally important. The accomplishment could have ranged from saving many lives to averting huge catastrophes. In some military units after their missions are successively accomplished, many of their personnel exclaim that phrase to each other.

One such unit I served in was the Fleet Readiness Action Group, or FRAG. FRAG was a highly select unit of U.S. Navy and Marine Corps active duty and civilian status military personnel. My particular assignment was in the early to mid-1980s was to a unit in Pensacola, Florida, and the unit strength averaged about forty-five personnel. Our mission was to assist our unit's assigned Navy and Marine Corps squadrons in keeping as many as possible of their aircraft in an 'up status' (flight ready). Our rendered assistance was both supply and technical in nature.

Three of the walls of our main operations office were actually aircraft status boards, with aircraft information extended from floor to ceiling. Thus, all assigned unit personnel were required having security clearances for office entry. The office had no windows, its doors were cyber-locked, and we all maintained our own unit field day (office cleanliness). No one entered our office without proper authorization.

FRAG personnel expedited aircraft components through the various phases of depot repair there at the Naval Aviation Depot. There were many times the expediting process was so 'hot' for a 'down' aircraft in the fleet somewhere in the world, that the FRAG agent would personally carry components from one repair shop to another. Other times, the FRAG agent would initiate back-rob orders for obtaining or removing components from aircraft in the depot repair process and shipping them to the fleet. Many times, the FRAG agents would actually take their own tools, enter the crash-damage yard, and dismantle parts from damaged aircraft stricken from the military's operational inventory. In all these efforts, FRAG personnel kept many aircraft flying more critical missions in the fleet around the world than ever before the FRAG units existed. At the end of our day, when leaving the office, we would look at our Aircraft Status Boards on our walls and proudly exclaim to one another, 'We saved the day!'

A few years later, I was assigned to an Air Force Special Operations Squadron as an Aerial Gunner. After successful tactical gunnery missions, the flight crew would be celebrating on the return flight to home base. Cheers could be heard crackling over the headsets of objectives accomplished, end goals achieved, friendlies aided, hostiles taken out, hostiles suppressed, sites dusted off, clean sweeps obtained, aggressors erased, targets neutralized, and the list went on and on. Intermingled in with all our cheers and hurrahs, ovations at the end of the day always came that 'we saved the day!'

Interesting comparison, isn't it? One scenario is working toward sustaining a weapon system, and another scenario is working toward total destruction of weapon systems. Yet, in both circumstances, rounds of saving the day were expressed.

So just what were the days saved from? In one state of affairs, the day was saved, in a sense, for a more secure tomorrow by keeping peace. In another state of affairs, the day was saved, in a sense, for a more secure tomorrow by removing peace from the enemy. One scenario shows force for reserving the peace, and the other scenario actually expends a show of force as sacrificial payment for winning precious freedoms. Yet in both scenarios, 'the day was saved.'

So then, how will it be at life's end in 'saving the day' with its many applications in life? Will we face a record which shows we were somehow satisfied in just keeping things in peaceful reserve? Or will we face a record showing we were ***proactive*** and even ***sacrificial*** in doing what was right and winning freedoms? Just as the ole cliché is of 'Freedom isn't free,' so also goes, 'Saving the day always has a price to pay.'

End Credits

Have you ever done something you were never given proper credit for? Yes, we have all fallen victim to those situations. What governs how we live out those events, though, is the way we deal with the 'end credits' of those scenarios.

End credits are most commonly known as those listings of recognition passing in review at the end of theatrical productions or movies. Many viewers do not pause for viewing the end credits, and thus, the supporting staff and crew responsible for developing, assembling, and producing the movie go by largely unrecognized. Sadly, too many times it is only the main actors in the cast who are recognized and remembered for their work in the movie. Such is similar in many areas of life. Far too many times, people fail in properly recognizing the end credits in the lives of other people's endeavors.

There was once a time while I was assigned to the FRAG mentioned in the previous entry, that a serious incident of 'end credit piracy' occurred. Select details and timelines of the incident will be sanitized so as not bringing embarrassment to certain agencies.

For several months one year, a certain aircraft oil pressure indicator became in short supply in one military supply system. This shortage resulted in the downing, or grounding, of many aircraft of certain TMS around the world. This rippled into the effect of limited mission capable aircraft, and then for mission flight sorties. The end result became partial loss of mission readiness for one facet of that military service.

All of the fleet squadrons I supported were contacting me on a daily basis pleading for availability of those oil pressure indicators. After months of this scenario and mission readiness failing to a critical level,

I began 'thinking outside the box' as the saying goes. Across the hall from the FRAG Office was one of base engineering technical libraries, and into its files I went.

Within those files, I dug up the technical drawings for the oil pressure indicator the fleet so badly needed. Oddly, in those drawings was found two other oil pressure indicators with the same part number, but different dash numbers. Further research revealed one of those other dash numbered indicators was identical to the one my service needed, with one exception – the external canister was one-half inch longer.

Having been a former Aircraft Electrician, worked on that TMS aircraft many months, and removed and replaced that particular indicator many times, I was very well acquainted with its location on the instrument panel in the aircraft cockpit. It was my preliminary assessment the other dash numbered oil pressure indicator could be a viable substitute for the one my service desperately needed.

Going to the FRAG supply computer terminal, I found the national stock number (NSN) for the particular dash part number of the oil pressure indicator I'd discovered on the technical drawings. Then, bypassing my military service's supply system, I began researching the other military service's supply systems. Another service had over 400 assets (indicators) in 'A Condition' (ready for installation) in their supply depots. Their low reserve limit was 200, and thus, if authority was granted by the Item Manager and Cognizant Engineer of that indicator for both military services, my service could 'borrow' up to 200 of them.

At the time, my service was in need of about 125 'A Condition' assets. Therefore, our critical fleet mission needs would not degrade the criticality of the supply for mission needs of the other service. It would be a win-win scenario in my technically non-engineering opinion. To my supervisor my findings and recommendations were immediately brought. After hearing my explanation, he immediately rushed me into our duty truck and drove me to the base engineering office.

The next afternoon, I received a phone call from the technical branch of the supply office of my military service. The representative informed me their office and our base's engineering office had reviewed my findings and recommendations. Everyone fully concurred and subsequently the other military service's supply channels had been contact for support and approval of 'borrowed' assets. Support and approval had been granted and a Technical Bulletin was being drafted that afternoon allowing substitution of the alternate part numbered oil pressure indicator I'd discovered.

The representative asked me to wait on the line for a moment because someone else wanted to speak with me. The next individual was one of the chief engineers for the project. Both representatives expressed their sincere appreciation for my diligent effort with attention to meticulous detail, without which this project would not have been possible. Additionally, my efforts directly resulted in a rapid change of mission critical flight status for many squadrons around the world for my military service. Within a matter of only about two days, literally several dozens of mission critical flights were undertaken again, and things were well on their way to becoming normal – well, that is for the U.S. military setting things straight in harm's way.

About a week later, in response to all this, I received a very nice letter of appreciation from my commanding officer. Subsequently, weeks later some of the squadrons I supported sent me letters of appreciation signed by their commanding officers. However, about a month later, rumors began flying around about a significant Beneficial Suggestion (BE) being awarded at the supply office of my military service. The BE was for an alternate oil pressure indicator, and was being jointly awarded to the two men who talked with me on the phone about a month prior.

Immediately, my supervisor and I placed phone calls to both of those individuals, and no responses were returned. Contact calls were made for several days and no response calls to my supervisor and I never were returned. There was a very informative media article in the next quarterly military magazine generated by the command of which that military supply unit was attached. The article contained information regarding how those two individuals who had talked with me on the phone, had researched and resolved the supply and logistics problem for the much needed oil pressure indicator for their service. The media article went on mentioning many had accolades from their commands for unselfishly contributing to their country and the support of the military services during a critical time of needed defense. One of the most eye popping catchers was the article line mentioning the BE the individuals were awarded and shared along with a photo. The photo was of them and a large representative check. The check was in the monetary four digit range (it was indeed very significant, but I'll not mention the amount).

For me, though, I was in the throes of resolving the problem 'in the trenches' and placing backorder requisitions for my squadrons to be first in priority for shipment of those 'A condition' assets when they became approved and available. For me, I had downed aircraft all over the world needing to be placed in mission-critical status – and that was my number one PRIORIOTY. For me, I was not afforded the luxury of time to submit any paper documentation for my discovery.

Saving the day for national security and doing my job cost me significant notoriety, not to mention several thousand dollars. However, saving the day for national security and doing my job won me the favor of many in the field. It may never be known for sure the full extent the blanket of protection those mission ready aircraft helped to provide for freedom in the far corners of the earth.

Such are the REAL end credits; knowing one did their job and made a REAL difference in the world. Real end credits make a better and safer world; they don't just line the pockets of its characters.

Flying John Wayne

It is fairly obvious that in some aircraft, there are no restroom facilities, a.k.a. bathrooms, latrines, toilets. Some of those glorious military flying machines, however, do have limited 'necessariums' of which the modified urinal is about the extent of the restroom accommodations. In my Aerial Gunner flying days on the AC-130 'Spectre Gunship, we flew long hours with our missions, and thus, our aircraft were equipped with a modified urinal.

For the other natural physiological relief of waste, well, as Clint Eastwood always said in the movie *Heartbreak Ridge*, 'a good soldier has to improvise.' Improvising for this urgent necessity relief of life usually consisted of a plastic bag lined around an empty ammunition box (ammo can). By the way, this was in the days before serious considerations regarding hazardous waste control. So if any environmental types are reading this book, take a chill pill.

Privacy for necessarium visits was not the best in the world either (or rather, in the air guess it could be said!), and was most always behind the stealth of a machine gun flak shield. After the necessarium visit, it became the sole responsibility of the aviator for his 'package.' Package discarding usually was in the central garage can in the aft cabin, and that aviator was required for ensuring his dropped package was a secured one.

Most everyone knows there are always deviations to everything, and so it was with this story.

One of our fellow Aerial Gunners was a very humorous individual. Jon greatly enjoyed joking with everyone about anything when possible. Jon, though, was a top-notch gunner and always conducted his duties professionally well. Jon was a strikingly well-built aviator. He was a six-foot-four, solidly chiseled specimen of a Special Operations military man. His demeanor earned him the nickname John Wayne from his flying comrades.

One night, Jon was on a tactical gunnery mission. The aircraft had completed its mission; the crew had inhaled a quick in-flight meal and then proceeded with engaging in a stealth exercise. During the stealth exercise, the IO (illuminator operator) is literally hanging halfway off the back-loading ramp watching for threats. When the stealth exercise was completed, the crew decided on flying low over the range until exiting it. The IOs sometimes would sit on the edge of the back ramp (tethered of course). Thus, the appearance was similar to riding the tailgate of a truck.

Sometimes we gunners would ride along with the IOs in tailgate fashion with a second tether. Jon, a.k.a. John Wayne, was quite fond of those aerial tailgate rides. One particular night, Jon decided on tossing his 'necessarium package' out the back of aircraft before taking his aerial tailgate position.

That smoothly planned tossing maneuver was executed well, but did not end up well.

The wind slipstream from both sides of the aircraft meets at the tail section about where the back-loading ramp meets the tail section when in the up-and-close position. Just aft of that area, an antenna protrudes downward from the tail section. Jon tossed the necessarium package clear of the back ramp, but just happened to 'land' dead center in middle of where both wind slipstreams join. The resulting turbulence from both slipstreams ejected the package backward and upward, and it impacted the antenna. The impact ripped open the package, and the slipstream wind 'sprayed' its contents into the aft aircraft cabin. Jon and the IO were the first aviators impacted. Four other gunners were in the immediate area and became secondarily impacted. The debris also impacted some areas of the aft aircraft cabin.

Six aviators stood totally stunned, looking at each other with splashes of excrement on them! In just a few seconds, all had gone from 'saving the day' from the aggressors to what a literal mess they had found themselves in! Their tailgate ride back to home base had just been transposed into a clean-up detail before landing (in aviation, if you make a mess, you clean up your own mess). So, to work the soiled crew went, cleaning up the mess.

But some stories do travel, like birds with wings.

When the aircraft and crew landed, not only did the ground crew sense some slight telltale evidence, but the story of Flying John Wayne had reached home base long before Jon arrived. At the Mission Debrief, the Debriefing Officer made amusing comments about dropping unauthorized hazardous waste packages. The Intelligence Officer even made comments about becoming innovative with dropping necessarium packages over the target after running out of ammo. Numerous other entertaining comments were projected about Jon's necessarium package event.

Needless to say, that event was never lived down. It became a source of comedy for many months and was never forgotten. Training classes mentioned the event. Inspections recalled the incident. Other organizations heard of the occurrence and, applying their own idioms, launched expanded versions. Those stories grew into tales (even though humorous) for being cautious in whatever actions pilots and crew might find themselves entangled in.

In general, though, with this somewhat sensible story, an analogy could be gathered from it for living a rational life. Be careful where unpleasant words, hateful opinions, and disgust are ejected. Whatever is in life that is offensive in nature, retain it carefully and deposit it where best manageable. If it is not, it just might come back and soil not only one's face, but one's life and those innocently in the surroundings. There is one very outstanding life lesson embedded within this story:

Aftermath's clean up of carelessness is always expensive, and often in more ways than ever thought possible.

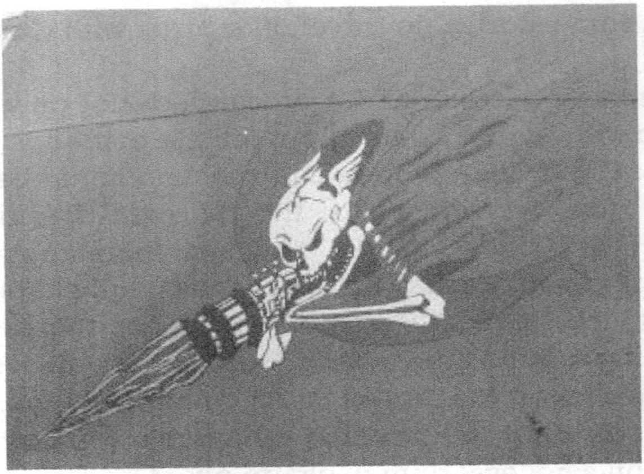

A hand painted Grim Reaper Machine Gunner image. These image decals were generally painted on the aircraft fuselage, and left side of the cockpit. This particular image decal was painted on an AC-130 Spectre Gunship which the author once was an aircrew member.

OMIAS

OMIAS, though sounding somewhat strange, is a well-known acronym in the United States Navy. The first time its sound ever came to my ears from the smiling mouth of my shop chief, I thought he was joking with me. He wasn't though. My request to be a 'backseater' in the squadron I was assigned to had been approved. The shop chief at morning quarters one day announced to everyone that Petty Officer Adams was going to physiological training and would receive his OMIAS card. The old-timers cheered, and the newer guys along with me had no clue what he was talking about.

Over the years, many believed OMIAS was an acronym, which all governments and militaries are very good with having within their operative systems. Over the years, some could never decipher the strange acronym. Thus, its readily conceived usage was quite plaintive involving the usual military humor. However, an in-depth investigation into the U.S. Naval Air Training Command will reveal that OMIAS actually is an existing acronym; it stands for the Order of Military Instantaneous Acceleration Society.

In the military, all aviators attend some form of Aviation Physiological Training (ATP). U.S. Navy Aviators first begin as Naval Aviation Cadets and attend the Naval Air Training Command's version of ATP. A short segment of the ATP can be completed by Navy and Marine Corps personnel who desire flying as a 'backseater' in jet aircraft. Some backseaters fly as Observers for a wide variety of select mission reasons. After completion of the required segment of the Navy's ATP, an OMIAS card is given to the graduate for the 'privilege' of flying backseat in jet aircraft.

The OMIAS card is a wallet-sized card with an image of a Naval Air Cadet dragging his worn-out posterior.

His worn-out posterior is due to it being 'kicked' with the force of acceleration from an ejection seat. ATP students undergo not only rigid classroom training but also actual training in attitude chambers and being shot upward in ejection seats. Graduate ATP students in the U.S. Navy and Marine Corps proudly carry their OMIAS cards. If not a Naval or Marine Corps Aviator, the bearers of an OMIAS card could perform duties as Observers, and they could also catch 'hops' from one destination to another—space available and mission pending of course.

Upon graduation from 'OMIAS training,' I was placed in 'backseater' status in the Squadron Duty Operation Office. A few days later, my Division Officer radioed down to my shop at about 0830 for Adams to get his flight gear, get suited up, and report to tail number 23 at 1000 hours for a Check Ride. All that jargon meant was for me to get my flight suit on, get it operationally checked, and be on the flight line at aircraft number 23 at 10:00 AM for an orientation flight as a backseater.

Incidentally, that aircraft had just completed a Phase Check, meaning, us maintenance technicians had partially torn it apart, accomplished some needed repairs, reassembled it, and then accomplished ground run-up checks. This flight was actually its first Flight Check, and as a backseater, I would be observing and recording aft cockpit instrument readings. An interesting item about this particular aircraft was that I had personally replaced all the explosive cartridges in both fore and aft ejection seats. This included both seat catapult rockets being replaced by a team of two other sailors and myself.

My shop chief released me for flying status about fifteen minutes later. While exiting our shop, he extended a firm handshake and, with his usual twisted smile, said, 'Go get 'em, Killer.' That was the first time I ever heard that nickname called out to me (and it would never be the last). Upon entering the Aviator's

Equipment Shop, I discovered the shop's Petty Officer in charge that day had already heard of my 'going up' and engaged me in a conversation while my flight gear was being suited up. Afterward, the line shack shuttled me with an NC-5 Jeep to the aircraft.

At 0945 hours, I jumped off the Jeep as it pulled up to the aircraft. My Division Officer (a Marine Corps Major) had already arrived and had just stowed his gear. He instructed me to do the same while the flight line crew began preparing the plane for launch. The major instructed me to walk with him as he proceeded with his preflight of the aircraft. He stated of understanding my having previously been a Plane Captain, and therefore, he wanted both sets of our eyes on the preflight.

When it came time for being strapped in the seat, the major just looked at me and said, 'Climb in, Adams. You're also a seat mech, so I don't have to babysit you.'

And so I climbed in, and the ground crewman assisted with the usual completion of strapping into the ejection seat. Sitting in that seat, the one I had replaced the explosive cartridges in, the thought came to me, 'If something goes wrong, the major and I may have to depend on my work for an egress outta a bad situation.' There is nothing like making sure one does a job right more than knowing someday you WILL depend on what you have accomplished – especially if the task may save your life.

Minutes later, the aircraft engines fired off, all run-up preflights were completed, and we were taxiing down the tarmac toward the runway. On the taxiway, the major briefed me on what he wanted from me. Of course this had been done previously, but he wanted a quick refresher anyway. Within minutes, the aircraft was airborne, landing gear retracted, and off we headed to a safe flight level over the Gulf of Mexico for checking aircraft performance.

The morning was proving to be a bright sunny fall day with hardly any clouds in the sky. The near crystalline-clear sky produced great visibility of probably around fifty miles, making the environment just right for flying. The only small alarm coming to us during the flight check was the number two generator delayed coming back on line twice after the number two engine air starts. After making final notations in our aircraft flight check logs, the pilot's voice came cracking through my headset, 'Hey, Adams, the skies look great. Wanna go for some real flying?'

My answer came back a quick, 'Yes, sir, you bet!'

Away the aircraft turned and banked out toward the sea, and the pilot's voice came through, stating, 'We're gonna head out to a gunnery range and do some acrobatics, ya game?'

And again, my answer came back with a quick, 'Yes, sir, you bet!'

A few minutes later, we were over the range and we leveled off at about three thousand feet. The pilot's voice came through my headset saying what maneuvers we would be doing, and away we'd go into them. We accomplished a few rolls, spins, banks, slides, near stalls, falls, and pull-outs.

After a few minutes, I heard the words, 'Adams, take the stick.'

'What, sir?' came my immediate reply.

'Adams, take that joystick in front of you and roll right and left,' the pilot replied.

'Roger that, sir,' was my reply, and so I did, and the feeling was great. That became the first time of many other times for 'limited aircraft maneuverability practice' as it was called. On the return leg of that sortie, however, my euphoria would meet sudden gloom.

Just before making landfall, the pilot stated he wanted to do a barrel roll and asked if I was game, and I stated of course. Suddenly, the aircraft joystick came back toward me and the aircraft nose immediately rose upward. My head went back into the seat headrest as with all the rapid acceleration climbs. The canopy view began changing as with the other times while the anti-G suit intermittently engaged for preventing fainting. This time, though, the beach area in the distance to one side was inverted and to the top of the canopy. What a trip all of this experience was!

Then the beach/sea view area came slowly in front of me and below me again. Wow, this was great! About the time the aircraft leveled out and accelerated into straight and level flight, my stomach felt as if it turned completely over inside. Then, without warning, the sensation of vomiting suddenly appeared in my throat! Reaching for a barf bag in my left flight suit leg pocket, the bending motion caused intense vomiting sensation. My next choice was reaching for my helmet bag in the map case compartment.

While unsnapping the case top, I upchucked into my oxygen mask! The exhalation valve at the mask bottom opened, and nearly all of the upchuck streamed down onto my flight suit and began filtering around the ejection seat pan. By this time we were nearing our final approach, and I squawked the pilot, 'Sir, I've upchucked.'

'Adams, for God's sake, do not upchuck in your O_2 mask' squawked back the major. 'Repeat, do not upchuck in your O_2 mask! Remove it now! I'll get us down in 'bout three to five!'

I didn't have the heart for telling the major it had already been accomplished. About ten minutes later, we were taxiing down the runway, the canopy was open, and fresh air was healing a sick stomach. Once the aircraft was spotted and parked by its Plane Captain, I came climbing out and told him I'd be back to clean up the mess. He just smiled. Walking across the tarmac, the major explained that if I'd eaten any serious greasy foods that morning, and a spark occurred in the mask when squawking the headset, an explosion could occur. This is because the greasy foods supply the hydrocarbons as a fuel source for combustion with a spark and the oxygen inside the O_2 mask.

When telling the major I had eaten a large breakfast of eggs, bacon, and gravy and what really happened in the aircraft – he stopped, rolled his eyes upward, and stated we were two very lucky men that morning. Then he asked if that episode was going to discourage me from flying. My response was a quick, 'Of course not, sir!' And so my first flying experience as a U.S. Navy backseater was a 'limited' success. However, that experience did not hinder me from becoming involved with other aviation ventures in my military career. A few years later, I would become an Air Force Aerial Gunner with a Special Operations Squadron. Again, I'd become bounced all over the sky, but this time through bullets, smoke, and many other things.

We've all heard the old cliché, 'If at first you don't succeed, then try again.' With select U.S. Navy Airman, they carry OMIAS cards for proving it – just ask them about it.

OMIAS Card
and some of the author's photos

Where Rubber Meets the Road

As a newly commissioned Second Lieutenant in the U.S. Army during the mid-1980s, Lieutenant Thomas Rogers had the distinct privilege of serving in a forward stationed logistics unit in Great Britain; and with the 47th Area Support Group (ASG) located at Royal Air Force (RAF) Base Burtonwood, England.

The 47th ASG was the active duty ground unit which ensured advanced preparations were being developed and implemented for a reserve unit to be deployed in the event of a war with the Soviet Block. As such, the 47th ASG was staffed at ten percent of normal manpower manning levels for this specific tasking. The 47th ASG was also the senior U.S. Army Command in the United Kingdom. This meant that all other U.S. Army units stationed in the United Kingdom reported to them and were under their command and control.

Within the 47th ASG, Lieutenant Rogers was the Operations Officer and reported to a major who subsequently reported to a colonel. That particular colonel had his sights steadfastly set on becoming a general at any cost. Consequently, the colonel decided to change the operating relationship between the 47th ASG and the reserve unit. The original plan was for the reserve unit deploying to RAF Burtonwood and absorbing the 47th ASG. The 47th ASG Commander (the colonel) developed an alternative plan for changing that relationship.

The colonel successfully obtained approval from higher headquarters for his new plan specifying that the reserve unit would relocate to RAF Burtonwood as originally planned, but would be absorbed by, and commanded by, the 47th ASG. As a result, the 47th ASG's small ten percent staff was left with the task of developing a new strategic mission plan, and coordinating it with all impacted U.S. Army Commands, as well as with the host nation's military command structure. The new mission became a daunting task. Most of the mission implementation responsibility and planning fell on the Operations Officer, Lieutenant Rogers, and two of his non-commissioned officers (NCOs). His two NCOs were a Staff Sergeant and a Sergeant First Class.

When the time came for annual evaluations, which determined all officers and NCOs promotion potential, the colonel was not pleased with the progress levels of his plan. The major called Lieutenant Rogers into his office and advised him of the colonel's high degree of dissatisfaction. He was further informed that someone had to take the blame for the perceived lack of adequate progress by the 47th ASG. Thus, Lieutenant Rogers was ultimately given a choice: (1) lay the blame on one or both of his subordinates and secure his own career advancement, or (2) give good evaluations to both of his subordinates, which would then result in him assuming the blame.

Lieutenant Rogers, a very junior officer barely and out of college, just could not believe he was left with only these two choices, but he was. Yet, he had been thrust into a situation most officers would never have to face, and if they did, it would have been as a senior officer (major or higher). Because he was well aware his subordinates were very diligent in carrying out their assigned duties, were working as hard as could reasonably be expected, he chose to give them the good evaluations they deserved.

This ***morally right decision*** effectively ended any possibility of Lieutenant Rogers making the army a career. However, that ***morally right decision*** preserved his personal integrity and allowed him to illustrate a Biblical principle to the colonel and the other senior officers; that principle is most commonly known as the Golden Rule. Christ Jesus expounded on the Golden

Rule principle in Matthew 7:12, *"Therefore, whatever you want men to do to you, do also to them, for this is the Law and the Prophets."*

Lieutenant Rogers received an average annual evaluation; one that is deemed a middle of the road appraisal for officers. Such evaluations are considered the 'kiss of death' for an officer, which indicates such officers are not promotion-worthy. Even though the details his superiors used for justifying their evaluation were drummed up, Lieutenant Rogers was never able to successfully counteract them. While Lieutenant Rogers still does not believe he should have had to make such a career-ending decision so early in his military profession, he never regretted obeying the Word of the Lord.

The Lord God, though, does have rewards for His people who obey His Word and follow His principles.

Not long after Lieutenant Rogers received an Honorable Discharge from the U.S. Army, he entered U.S. Government Service as an Air Force Civilian. Entering government service, he carried his army seniority with him and his army experiences as well. Mr. Rogers ended up serving the USA again working as a Logistics Specialist at the Wright-Patterson Air Force Base in Dayton, Ohio. That air base was where I became acquainted with Tom in 1993.

One of Tom's duty assignments became the same joint service unit I was assigned to, the Joint Depot Maintenance Activities Group (JDMAG). Within JDMAG, Tom was assigned to the Joint Technology Exchange Group (JTEG); and, at the time, I was assigned to the Joint Depot Environmental Panel (JDEP). The JTEG and JDEP were the elements comprising one division within JDMAG, and managed select technology issues for all of the U.S. military.

Through the years, Tom has been promoted to the U.S. Air Force Civilian grade level equivalent to that of a military lieutenant colonel. Along that pathway, Tom entered the ministry, became a Reverend, and, for a while, was even a church pastor. Along the way, Tom taught many church classes and held many offices in church. During those years, Tom successfully provided for his family of four. Tom's army career may have been sniped from his grasp at a young age, but because of his devotion to God and His principles, God favored him, and made a new pathway for Tom's life.

God spoke in Isaiah 43:18-19, saying, *"Do not remember the former things, nor consider the things of old. Behold, I will do a new thing, now it shall spring forth; shall you not know it? I will even make a road in the wilderness, and rivers in the desert".* God made a road of new beginnings and a river of refreshing in the wilderness and desert places of Tom's life generated by his army superiors.

In bringing this story to a close, there is more to the story with Tom's two superiors who were instrumental in railroading his active duty career – the major and the colonel. The major retired at that same rank; he never achieved another promotion. Interestingly, sometimes when we go along with the steam roller of evil plans, we can end up getting caught in its evil pathway as collateral damage. As for the colonel, he succeeded in his drive of becoming a general; he got his precious star. By employing less than honorable management practices, he achieved his destined goal to wear the rank of a military star on his uniform. However, not long after he retired with his beloved star, he died from cancer.

Evil is insidious and eats away at the linings of the soul. Once it breaks through the soul's seal, it can burst forth in destructive pathways most difficult to cease. Health issues can be just one of many of its menacing avenues of devastation.

At the time of scripting this book, Tom is of the comparable military rank of lieutenant colonel. He has but a few years remaining before retirement and thus, will most likely not achieve a comparable civilian grade level of general. With retirement becoming more defined on his horizon, he has other occupational opportunities and ministerial callings rising into view as his military career begins a sunset. As previously mentioned, though, he has well succeeded in many avenues of life, including another branch of government military service.

In looking back at Tom's beginning days in the U.S. Army, he was faced with a very stressful career ending situation; one which might be sometimes called 'where the rubber meets the road'. It is during those times that the principles of God are put to the test against the principles of the world. God's principles *will always prevail* – sooner, or later. *"Therefore keep the words of this covenant, and do them, that you may prosper in all that you do"* (Deuteronomy 29:9).

One other comparison is very interesting – that of the star. Tom became a minister, and thus, has the title of Reverend before his name. At one time he was the pastor of a church, and as a result, became a different star – one of God's 'stars.' There is a reference in the Bible where pastors have an analogy; of being referred to as 'stars' and is found in Revelation 2:1. The passage of Scripture is a rough one for that analogy, however, some scholars and teachers relate the star in that passage as the pastor of those churches mentioned.

All of the celebrity who's who stars in this life will never compare to the humble stars within God's Book of Life.

In all the years I have known Tom I can personally state that Tom is of rare moral fiber. He is of the caliber that stands for the right regardless of the circumstances. When everyone exits the off ramp for the easy way out, Tom makes the right turn even if that path seems impossible – just because it is the *right* thing to do. Tom's dedication to duty AND devotion to the Lord God will come shining through *in red, white, and blue*, in spite of the few who would tarnish it!

Within this story, Tom's dedication to duty, and his devotion to the Lord God was put on display for his superiors and subordinates alike – in both detailed inspections and examinations. However, those assessments which count the most are by the Chief Inspector – the Lord God. It is God that Tom will ultimately answer to for a *life evaluation* someday – and not anyone else. When that day comes, the events of this occurrence will shine again – from the Lord God's Book of Record, and *in red, white, and blue*! The Lord God rewards His people for doing what is right – both in this life, and the next. Reverend Thomas Rogers is living proof of the principles of God in action.

Thomas Rogers
Photo was taken before he
departed for England in 1982

Double-Time Sailors

After completion of boot camp in 1967, the U.S. Navy sent me to the Naval Air Technical Training Center (NATTC) in Memphis, Tennessee. The base was actually near the small town of Millington (then). When entering the second phase of technical training for the trade skill of Aircraft Structural Mechanic, the school staff appointed Class Leaders and Assistant Class Leaders for both Navy and Marine Corps trainees. For my class, the original selections were for me being the Assistant Class Leader. However, personal problems were encountered by our Class Leader and the amount of leave he was required taking set him back into another class. This action placed me as the new Class Leader for the sailors, and occurred about a month before the ending of our training program.

One of the Class Leader's responsibilities was marching the class to and from school every day; whether clear, rain, sleet, or snow. The length of our march was about a mile one-way; AND, we were never to be late for classes. Class Leaders were to march near the forward section of the GI column, call cadence, render all military courtesies for the column, and keep the column of GIs looking sharp at all times. If the Class Leaders did not, they were the ones to answer for any observed (or perceived) misconduct of the marching column. There was one difference, though; it was double-timing marching orders with the Navy and Marine columns of GIs. The Marines were permitting marching in double-time, but the Navy was not permitted. We sailors all thought it was a ridiculous policy, for on other Navy bases sailors were permitted marching in double-time, but not at NTTC Memphis (at that time).

Only a few days after assuming duties as the Navy Class Leader, there came a day when all the Navy and Marine trainees were detained longer than usual after training. A special inspection was coming up in a few days and everyone was pitching in a little extra with polishing up the training equipment for the inspection. When finally being released for falling out into formation at the end of the day, the realization struck us that we might not make it in time for evening chow. The Marines were not as concerned as the sailors were. The Marine chow hall was not the same distance as the Navy chow hall, and the Marines could double-time if they wanted to. Besides, the Marines were always double-timing around us sailors, and smugly smiling back at us as they ran by anyway.

Once we cleared the school grounds and entered onto the main road leading back to the barracks side of the base, we came up behind a marching column of Marines. They did not seem to be in any real hurry because they had time to get to their chow hall. Many of the sailors called out they did not want missing chow over being blocked by the Marines, and for them to move into the road. The Marines called back that we sailors were a bunch of wimps and always whined if we missed mommy's milk. About that time the Marine Class Leader turned backwards and smiled at me, but never said a word.

That was about the time when I'd had enough, and the Marine Class Leader pushed my final buttons.

Calling to the sailors in my column, I gave the marching order of right oblique and when we were in the middle of the road, then the order for a left oblique, and then the order, 'Double-time, on the stride, mark'. We passed that column of Marines with ease while many of them hailed obscenities toward us. Once clearing them, I called the reverse marching orders for bringing us back to the road's edge, but I did not bring us back from double-time until we came to the main road before the barracks area. It was there we were spotted by an Admiral's Staff Car. It was definitely a staff car because, as it approached us, the admiral's was flag displayed and I saluted the car – while we were illegally engaged in double-time. Some of the sailors in my column called out, 'Adams, we're in trouble now.' And to their calls I replied, 'No, I am, not ya all.'

At class formation for the sailors the next morning, one of the school administrators was there. After a few administrative comments, he dismissed everyone except me. I was then directed to report to the senior school administrator's office. Later, some of the guys in the class told me they made remarks that I was being sent to the principal's office. Once in the office, a Navy Chief told me to get ready for a royal chewing out. Asking why, his eyes nearly popped as he said, 'Adams, you ain't stupid. You know what you did yesterday with breaking the rules.' Then he opened the door to the Administrative Officer's office, we both saluted, and the Chief exited while closing the door.

The Navy Lieutenant remained seated and had me to remain at attention while he proceeded to read the base no double-time policy for Navy personnel in formation. Then he asked if I had broken that policy and when. Replying that I had and it was the day before, he wanted to know why would I do such a thing? My explanation was that our class had been detained later than usual, and we were very possibly going to be late for evening chow. Therefore, the circumstance placed me in more of a leadership position than usual.

His face went from snarling to questioning, and then to a blank.

After recovering from the damage control my answer had struck, the lieutenant asked me to be at ease and explain the last part of my statement. My response was this: 'Sir, as Class Leader, I am responsible for many aspects of my men. One aspect is safely marching them to and from this school premises. In carrying out those duties to the best of my ability, I must see to it that they arrive for receipt of proper nourishment at their assigned meal times. If school functions detain us, and the Marines get in our way, then I am forced to make decisions which break ridiculous policies, sir.'

'You double-timed your men around a column of Marines?' the lieutenant asked.

'Yes sir' was my reply, 'and I'll do it again if placed in the same situation, or until relieved of my duties, sir.'

The lieutenant stood to his feet and smiling said, 'I was not told you double-timed your men past a file of Marines. That demonstrated some unusually unique initiative, Adams. I will smooth this issue over for you with the admiral. By the way, be giving some serious thought as to where you'd like your next duty station to be. You are dismissed, and have a great day.'

Saluting and responding with a 'thank you sir,' I did an about face and exited the office door. The Chief cast me a puzzled look as the lieutenant called him into his office. That double-time situation was never mentioned again. And, the smiles of all of the class instructors seemed to be somewhat slightly different toward me after that.

A few days later, we sailors filled out our 'dream sheets' for upcoming transfer orders. My first choice was any Helicopter Gunner, Vietnam; second choice was any Search and Rescue, Alaska; and the third choice was any Helicopter Squadron, Pensacola, Florida. About a week before graduation our orders came in, and my orders was for Helicopter Squadron Eight, Pensacola, Florida. I've always thought that lieutenant somehow had something to do with those orders. Why? Because he asked me once before leaving the school how I liked my orders.

So, initiative for doing the right thing for those who depend on you can pay off in more ways than just one, even if some stupid policies must be broken – well...most of the time anyway.

Court Martial Offense

"That's just their same ole 'Mickey Mouse' stuff going on again." Every individual who has ever served any capacity in the military has heard that phrase, and has most likely uttered it at least once. There are other 'quaint' little phrases GIs use for addressing the seemingly ridiculous situations which always occur within military life. For this book, though, those colorful axioms will not be mentioned. Mickey Mouse will do; for it implies very well the petty attention to detail applied in areas which the average GIs know full well need not apply. The following story is *just* <u>one</u> of those ridiculous applications.

Two of my fellow unit aircrewman and I were sent to several days of training. The base location and type of training will not be disclosed. The training was open to all four services and select agencies of the U.S. government. On the very first day of training the orientation briefing was given to over fifty students in the morning session of one classroom. In that session were aviators from all branches of the military, and a couple federal government agencies. As was directed by school policy, all students were dressed in their appropriate agency flight suits.

My fellow unit aviator comrades and I were seated about the middle of the rows, and about three-fourths of the way back from the classroom front. The instructor was up front delivering the lecture while three of his assistants patrolled the room with snake eyes. The instructors and assistants were dressed in black uniforms (this was for a unique reason for this specific training program). Those roving assistants ended up walking through every row of seats in the classroom. They ended up tapping us three on the shoulder as well as a few others. At the first break, we three were given an official looking slip of paper. It stated our names and that we were required reporting to the school administrator's office at the close of training that day. During the break, we talked among ourselves and with some of the 'identified' others, wondering what this meeting could be about. No one had any clues.

But then, when returning to class, it hit me; haircut policy, what else!

My rationale was this: all of the GIs singled out were from high-profile units, rather than low-profile units. Therefore, someone in the administration section of this training center was of the opinion that elite military units should outstandingly 'shine' in every aspect of their performance and appearance. That opinion included their haircuts (and 'high and tight' haircuts, if possible). So, I immediately began noting certain grooming aspects of other aviators in the class. In the course of that day, my notebook contained name, rank, and unit, of several GIs whose hair was notably longer than my two comrades and I – and were not gigged. The other GIs who were gigged in the class were not my concern (I was the senior ranking NCO from our unit in this class at the time).

At the close of the day, we 'criminal' offenders found our way to the administrator's office. A Master Sergeant seated us (seven total) outside his office and told us the nature of our offense. As I'd suspected, it was haircut policy violations. The GIs began an uproar and the sergeant ordered us to shut our mouths because he was doing as ordered himself. He then escorted us into the administrator's office. We saluted the captain in unison, and were then told to stand at ease.

The captain related we had been identified by his instructors as those needing haircuts, and therefore, we WOULD receive them before receiving a certificate of completion for training – or else. With all the intense schedule of the training syllabus the only opportunity for a haircut would be a couple of hours before taking the final examination on the last day of training. There would be no discussion of the matter and we were dismissed. The Master Sergeant saluted the captain, which signaled us offending captives that we must follow suit. We saluted the captain, he returned the salute, we were dismissed and did an about-face, and the Master Sergeant escorted us out of the administrator's office. Once in the Master Sergeant's office again, another uproar of complaints exploded from us all. The Master Sergeant raised his hands and stated it was not his call, but the administrator's, we had required homework projects for the next day in the field, and we were dismissed.

The march to our barracks was one mixed with anger. Everyone stated their commands had inspected them before departure and had been pleased with them. Therefore, how can this school command get by with apparent harassment, and how will this action really be handled? My advice for the GIs was that for the time, we all had to concentrate our frustrations on our field projects, and excel to the top our class, get those beloved haircuts, and take the examinations. If we could do this, our elevated status within the class will enable us to gain respect; of ourselves, of the class, from the school administrators, and for our units. Then, when we return to our units, we could file complaints through the official military channels if we wanted to do so.

All the GIs seemed to think the advice given them was good. All seven of us did, indeed, excel to the top of our class in that training course. In every exercise we seven GIs watched each other's backs and did whatever we could, above and beyond the norm, for making sure we seven excelled. Along the days of training, I gave each of the other six GIs my list of GIs who were identified as having longer hair than we had. Additionally, during the course of training, I made more notes as to unprofessional conduct and uniform appearance of those GIs with longer hair, and also those of the training instructors. That information was also shared among us seven offending GIs.

When the last day of class came, we seven GIs marched to the base barber shop. Within one hour, we got a precious military 'high-and-tight' and then marched to the school administrator's office for inspection before going to class for the final exam. After passing a beaming smile of an inspection, we were dismissed. When entering the classroom, everyone else was already seated and listening to the instructor rambling on with some sea story yarn. Once we 'magnificent seven' were seated, the final exams were passed out, instructions given, and the examination was commenced.

At the conclusion of the final examination, certification presentations, and administrative remarks, one final step remained. The students were given course evaluations to complete if they so desired. Many of the GIs had flights to catch and did not submit evaluations. However, several others of us GIs had later flights and stayed for completing the evaluations. Among those who stayed were all of the 'magnificent seven.' In fact, we seven were the last GIs to turn in evaluations and depart the classroom. All seven of us GIs took our time and completed the forms very *carefully* – especially me. I was the only one who had requested a couple pages of paper for use as supplemental information sheets.

What the other GIs documented on their forms was never revealed to me. On my evaluation, though, I documented the unprofessional and unfair slithering of the facility haircut police and their Mickey Mouse haircut policies, while failing to enforce sensible uniform policies. My evaluation documented the other GI students who wore longer hair than us seven who had been singled out. I further documented my unit inspecting us three for haircuts and other criteria one week prior to reporting – and us successfully passing inspection. My evaluation stated the training school was in violation of multi-service policy discrimination, and was apparently discriminating against elite units. There were some very unhappy staff faces when the magnificent seven turned in their evaluation forms that day.

While at morning section formation one morning about ten days after returning from training, my Section Lieutenant called out my name to stand fast after the section fell out from formation. After hearing variations of 'Sergeant Adams is in trouble now' and after everyone had cleared away, the lieutenant related that I was to report to the commander's office at 0900 and not to be one minute late. When asking why, the lieutenant just responded it had something to do with the last training school I'd attended.

When arriving at the commander's office, I saw standing outside at parade rest the other two GIs who had also attended the school along with me. A Military Police (MP) was also standing with them, but on the opposite side of the commander's office door. The MP looked at me and inquired if I was Sergeant Adams. Responding that I was, he instructed me to fall in at parade rest beside the other two GIs and be silent. About five minutes or so later the MP escorted us into the commander's office.

The commander was a full-bird colonel and was in command of our unit. When we first entered his office, readily could be seen was the fire spewing from his eyes. The MP along with us snapped to attention, salutes were rendered and returned, and the MP reported that the aviators in summons were present. The colonel dismissed the MP and he departed to just outside the door, closing it as he left. The colonel ordered us to sit, but not to get too comfortable for we had placed him in an uncomfortable situation, which needed answers.

The colonel went on explaining the commanding officer of the school we had just recently attended had written him a 'nice' long letter. The letter claimed we three aviators were three of the sorriest GIs to ever have come through the training facility during his assignment. Furthermore, we three GIs were prone to incite mischief, were very insubordinate, a discredit to our uniform, and the military service. Slamming the letter down on his desk, the colonel jumped to his feet and demanded an explanation of our actions as to what would result in such a 'confoundit' letter being written – not to mention the many negative phone calls he had received regarding the issue. Looking and pointing his finger straight at me, the colonel shouted, 'Sergeant Adams, you were the ranking NCO for this mess up, what happened? What is your report of this matter?'

With the colonel's direct commanding invitation, I began giving a truthful report of what occurred as he sat on the corner of his desk. After completing my rendition of what happened, the colonel's face began changing color. It began a slow burn into shades of red as he inquired of the other two GIs if they agreed with my report. After they quickly agreed, the colonel arose to his feet and we followed suit. He picked up the letter, waving it around, and with his face the shade near blood red, burst forth, 'Gentlemen, there's more to this here story than ya all are tellin' me or I'd never have received this here letter (waving it at us). I personally know the commander of this school and he woulda never written such a letter if ya all three had not done something to tick off his staff. You three have caused me to do some unwanted explainin' to my superiors. If ever I hear of ya all three causing another ruckus such as this again, trouble will be comin' ya alls way. Ya all are dismissed. MP come and collect these men outta here.'

We three saluted in unison, turned an about face, and the MP escorted us out into the hallway and then back to our respective duty sections. This issue was never brought up, mentioned, or discussed again, at least within that unit assignment. The letter our unit received in regard to us three GIs, and its resultant inquiry, did not sit well with me. Subsequently, some weeks later, I filed a Complaint with the Office of Inspector General (IG) which that training command was aligned beneath. The Complaint was very factual and did not contain any abrasive elements (in my opinion anyway). Several days later I received a letter from the IG's office stating they had received and logged in my Complaint, and would investigate it. Results of the Complaint were never received – by me anyway. In fact, this issue never came to light again until many years later.

Several years later (and even after leaving the military), I was going through some of my old military records. In one box were some papers obtained from this unit assignment when being separated from it. Among the papers was a large brown envelope which was still sealed. Opening the sealed envelope revealed an unpleasant surprise. In the mixture of official personnel papers was one form indicating the unit had recommended me for a Court Martial! Reading over the form further revealed my offense was Insubordination during the time of being temporarily assigned to training at that particular training base for a few days! What a surprise after all those years!

In the U.S. Navy I'd been arrested and incarcerated, and had also been to Captain's Mast pending a Court Martial. However, this was much different; there was in my hands an *officially signed recommendation* for a *Court Martial*, and it had been submitted by my own unit! Interestingly, there were no papers that could be found where higher authority had taken action on the recommendation. Evidently, since no negative actions had been taken against me, higher authority had disagreed with my unit.

It was not until drafting the manuscript of this book, and this particular story, that something occurred to me which never had before in regard to this story. Recall the earlier *Lost at Sea* story? Recall in that story of a

Chaplain Arthur Baron, who, when listening to the *Lost at Sea* story/testimonial, arrived at a very unique conclusion?

Chaplain Baron came up with the possibility that the crewmen responsible for loading the survival pack may have conceived a plan to place most of the defective equipment in that one pack (which became mine); then inform the rescue spotters to watch for the one trainee splashing most in the water. However, someone else may have had somewhat of a sadistic thought process, and placed all the remaining defective equipment in that one very same survival pack. The results being the rescue spotters would see someone ***desperately*** troubled in the water. What was probably not taken into consideration was the possibility of strong surface currents – one which rapidly separated me from everyone, and carried me far out to sea.

I had previously written, that in reality, satan could have orchestrated the survival pack events one day, and then the sea current events the next day. Thus, those events placed me directly in harm's way for becoming totally lost at sea with no survival equipment. Those events could have been orchestrated by evil, fostered by satan, in attempts for permanently taking me out of this life; and at that time, it would have been far out from sight of everyone.

While drafting the manuscript for this book, it occurred to me that my initial challenging of the school officials over haircut policy was taken as an *affront* to them, when actually my actions were not intended as rebellious. Therefore, it just may have been that **ALL of the defective survival equipment *actually WAS intended*** to have been placed in my survival pack. In doing such, my desperate struggles for survival were **intended for humbling me** and teaching me a lesson (as the saying goes).

However, the events of that day backfired on them all, resulted in a near fatality, and the colonel who was responsible for all of those events (and may also have had it in for us seven GIs) had to make a public apology to me. Thus, the colonel, who became infuriated over haircut policy and his staff's opinion of rebellion, actually became publically embarrassed. In order to counter the public embarrassment of himself, his staff, and his unit, he retaliated by setting out to write an inflammatory letter. That letter would be an attempt for tarnishing my flying and military career (or burning it as the saying goes).

Such are my opinions anyway; and the facts do appear in support of them (don't they?).

Even though there are seemingly too many useless Mickey Mouse activities going on at times within the military, most of the GIs overlook those ridiculous messes because their hearts are focused on serving their country and its ideals. The cores within most American GIs are stained with over two hundred years of *red, white, and blue* – and they cannot allow a few ridiculous practices to impair their vision of what those overlaying patriotic colors represent.

There is a tremendously important life lesson to be learned and gleaned from this near court martial story. Evil will go to the ends of the earth for taking us ALL down with any temporary opportunity toward permanent immobility. Nevertheless, beyond the horizons of today, higher authority will always attempt having the final say over everything of tomorrow; and that includes modifications to total disapprovals. It just may be that when this life is over we will see the many 'stories' in our lives where 'The Higher Authority' has ruled out evil demands; those of wanting to 'court martial' us all into the valley of death – and far more times than were ever realized.

"The name of the LORD is a strong tower; the righteous run to it and are safe" (Proverbs 18:10).

On Patrol

In 1975, my U.S. Army Unit was called in for patrolling and guarding an area devastated by Hurricane Eloise. The Fort Walton Beach to Panama City area of Florida was a total mess for several weeks. No electrical power, buildings blown down (and away), roads and streets caved in; the region was crippled for a long time with cars, trucks, and boats strewn and piled up everywhere.

The first contingent of my unit arrived late in the afternoon and we set up a Command Post in the Civic Center. We were informed our supplies were on the way. We thirty-plus soldiers were getting hungry. Some of us packed along some extra canteens of water and food from home, and shared what we had before getting ready to go out on the evening patrol. Just before dark a large shiny white truck with red crosses on it drove up in the parking lot corner of the Civic Center.

Three of us were directed to go and investigate its reason for parking there. We discovered the truck was setting up a distribution point for food and water. When the sides flew open its interior revealed stashes of packaged sandwiches and bottled drinks. The driver told us one sandwich and one drink for one dollar. All three of us promptly made a purchase and then reported our findings back to the unit captain. The Command Center subsequently emptied of its soldiers as the parking lot corner filled with GIs purchasing food and drink.

Morning came and no supplies, not even the chow truck. The unit supply trucks did not arrive that day. Those of us who were back at the Command Center wanted fresh water and some evening food. We found, though, the incoming soldiers from the night patrol had been fed fairly good; it was by the civilians in the area of their patrols. People had been out grilling breakfast that morning. There was no power, and everyone knew what was in the refrigerators and freezers would not last long. So, the people offered whatever they were grilling very generously to the passing soldiers. Those scenarios remained the entire time we soldiers we there on patrol.

By early evening the white truck with red crosses on it was still there selling food and drink. While we were standing in line, another truck came up and parked in the adjacent corner of the parking lot. It was brown in color and was not as new and large as the white truck. On its sides was painted in red the large words 'Salvation Army.' Two of my buddies and I looked at each other with raised eyebrows, and then I said, 'Let's check this out fellas.' We fell out of line and crossed over toward the Salvation Army Truck.

Arriving at the truck, the driver asked us why we were standing in line at the white truck. We finished telling him our hungry story about the time he had just popped open one side panel. As with the other truck, its interior revealed packaged sandwiches and bottled drinks. We asked how much, and the driver's mouth dropped in surprise saying, 'Why fellas, there is no cost. It's free.' In response we all three of us said, 'Na; Yer kiddin; Can't be.' The driver once again told us it was free. All he asked was just to limit one sandwich and one drink per soldier for each trip he would be making until our chow provisions arrived.

At this good news, we three whistled and yelled at our other buddies to come over and check out the free food. The white truck became deserted as the brown Salvation Army Truck became engulfed in olive drab green. Still, though, there remained a few civilians in line at the white truck. Our supply trucks arrived later that evening, and we had a great hot Army meal for dinner. The cook heard of what the Salvation Army truck driver did for us, and he carried him a personally prepared hot army meal. That Salvation Army truck driver also received an invitation to dine with us whenever he desired. Interesting was that the white truck with the red crosses on it moved on to another distribution point.

There are some spiritual life lessons here with this story. When the storms of life come, and they will come, they can knock all the power right out of our lives. Those storms can be so severe that the fabric of our lives can be torn totally apart. In all the residue from the wake of those storms, there may be friends who give us the leftovers from their lives. Their gifts are sometimes not ill intended at all; it may be all that they have. But too many times the leftovers are not sustaining enough for needs within. Those good gifts will only last for a while.

In our foraging through the aftermath of damages, rescue operations come by. Those rescue attempts may be structured programs, of which many are excellent, but they may come with price tags. Often times the invoice from those formalized remedies can completely deplete our resources. Then in waiting patiently, on the other side of the parking lot of our life, another Rescue Operation comes into view. His name is Christ Jesus, and His work is Heaven's Rescue and Salvage Operations. His salvage operation comes with no cost; it is free for the asking, limit one per person, per lifetime. Christ will be in that salvage operation for as long as it takes to rescue that lost person and refurbish them into complete redemption.

When the events of this true-life story happened to me, I did not immediately see the spiritual applications in it. The applications came a few months later when pondering over the events of that deployment. However, it was one of God's Imprints given to me for encouragement along my journey in life, and for later sharing its message with many others, and also the readers of this book.

Windows Platoon

Once in the late 1970s, my U.S. Army Unit was in Fort Lewis, Washington. We had come from spending several days in the field (woods actually), and some of us were in the back of the transporting deuce-n-half (army truck) discussing a tour of Mount Rainer over the weekend. Upon returning to base, one of the soldiers accomplished some 'recon' and came up with the possibility of a charter bus. However, he had to know right then if we wanted the charter because the charter line had another client on hold. All of us in the barracks who'd been discussing this quickly agreed and the recon soldier chartered a bus for about two days from then.

About thirty-six hours later, at 0800 hours in the morning, about sixty-some soldiers were standing in the barracks parking lot waiting on the charter bus. About 0815 hours this nice new looking Greyhound bus pulls along next to us, stops, and the driver jumps out. He looks at the manifest of men handed to him by our unit captain, who has also chosen to go with us. The driver's eyes quickly scanned over the page and then shifting his hat to the back of his head, asked all of us to gather around him for a moment.

The driver told us the basic day's agenda in the style of an army drill sergeant. We all began looking at one another rolling our eyes like 'here we go again.' As the driver's instructions finally began winding down, he changed into a new topic – the new bus. The driver began telling us the bus was a brand new one, and this charter was actually its run. He asked us to please be respectful of the new furnishings and police up after ourselves.

The driver went on relating the charter company had scheduled another bus for this run, but it had to be used for another commitment. Unscheduled maintenance problems were the reasons for changes of buses. Actually, he said, this brand new bus was the only one left in which to make this run, and management wanted to cancel it because it was going to be used for transporting GIs. At this, some moans and disgruntled looks emerged from the troops.

The driver then jumped onto the bus entrance step, raised his voice in drill sergeant style, and quickly regained control of the troops at the surprise of our captain. He called out that he was a retired army sergeant, he was very sympathetic toward any GI, and he assured his management the GIs of his run would treat his bus with great respect. He told his management he did not wish to disappoint the defenders of their country by canceling their tour. With that patriotic statement, the driver got a loud round of applause and many troop hooras.

When the cheering and hooras cascaded down, the driver jumped down from the bus entrance step shouting, "Men, on to Mount Rainer" as he waved us onboard. Once the bus was loaded with us army soldiers our Platoon Sergeant took final roll call, and then the bus driver stood up for some routine pre-departure bus safety instructions and tour comments. He also stressed again to please keep the bus clean for it was going to be used in two days for transporting some foreign dignitaries. Then he climbed in the driver's seat, and our mountain tour commenced.

About forty-five minutes later, we were near the end of an overpass and stopped in traffic. Sitting on the right side of the bus, the view from my seat was through the rear half of a large window. Sitting there I observed two young boys in a vacant lot across the overpass barricade. They were picking up stones and throwing them at seemingly whatever caught their attention. There came a time when one of them was standing and just tossing a stone up and down, searching for a target.

His eyes lowered on our bus, and then his head reset in our direction like radar. His mouth uttered something as his rocky partner spun around zeroing in on our bus.

My immediate thoughts were, 'Oh no, they are thinkin the wrong thoughts.' About that time, the soldier sitting next to me sees the same thing I am seeing. We both look at each other and say, 'Hope they really don't try that.' Those two boys begin tossing their stones up and down in a fashion that is obvious they are going to throw them. Then we both reach over the seat in front of us and get the attention of those two soldiers. They peer over and see what is happening and then we all yell out loudly 'Eyes right guys!'

About the time some of the guys on the right side of the bus looked right, two large stones impacted the window right next to me. Two resulting large cracks splintered themselves down toward the bottom of the window. The driver called out what was that, and several men came back with some boys hit the bus window with rocks. The driver just hung his head, shaking it back and forth. The two boys were nearly rolling on the ground with laughter at the stunt they had just pulled off. The gridlock traffic then eased, the bus began rolling, and the two boys waved us off in hysterical laughter.

Our captain jumped up, looked back at us and spouted out, 'Hey, wanna get them guys?' We all came back with, 'Yeah!' The captain told the bus driver to pull the bus off at the next right and head back, and so we did. About three blocks later we entered an area of the vacant lot. The boys were at the other end of it and turned around when they heard the bus. When they saw the bus, both of them dropped their stones and ran as fast as they could. With the bus relentlessly pursuing the sprinting boys, they entered an open field. The driver said from the heading they were taking, he knew about where they were going.

At the edge of the open field the bus came to a stop and the captain called out, 'Hey men, ready for some PT this mornin? Let's go get'em!' He told the driver to meet us on the other side, as sixty-some soldiers filed out of that bus and broke out running across that open field. As I filed out passing the driver, he had one huge grin on his face. None of us hesitated as we expelled ourselves from the bus. As soon as we hit the ground, it was off in running, and the Chase of Seattle was on that day.

Those two boys were ahead of us, but not for long. They entered a housing area on the opposite side of the open field, and so did we. That pursuit had to be a scene that morning; two teenage boys being chased by sixty-some soldiers down the street of a housing district. Turning a corner, one of them disappeared into a backyard. The other one ran into the front door of the house, slamming the door behind him. Just as the door slammed, sixty-some soldiers ended up on the front lawn of that house. Our captain knocked on the door and a man answered. You should have seen the look on that man's face. One would normally expect one or two visitors from the knock at the door, but not sixty-plus visitors!

The captain wanted to know if the man knew where his son was. About that time the bus pulled up in front of the house, with the cracked window showing. The question-filled man called out to his son, who was very red-faced and in tears. The bus driver had already called the incident in to his dispatcher, and subsequently within moments, the police arrived. The two boys were arrested to the dismay of their parents. The police thanked us for helping them. We resumed our tour for the day, and at the end of it gave the bus a spotless inside cleaning job. The driver was one appreciative man, thanking us for what we did.

Later the charter lines refunded our tour funds as a show of appreciation for our services to them. Our unit also received letters of appreciation for that episode.

The two teenage boys served incarceration time in the local Juvenile Detention Center. The families of the two boys were court ordered to pay the damages for the bus. I highly suspect those boys did not throw any more rocks at passing buses.

There are some lessons to be learned here in this true-life story. Keep the windows up on the bus of your life, for they keep the world out. If we go around with our windows down, the evil of this world will always throw in its trash. The windows of separation will remove us from openings to temptations. Then evil cannot find easy access into the vehicles of our lives. We also need to keep our spiritual fuel tank full and life's engine tuned up and running; being ready to go where the Captain of our Soul directs as essential. Regular Bible reading and studying, and praying will keep the spiritual fuel tank filled. When the world throws rocks at us, the Captain of our Soul, with His Salvation for us, will rise up and lead us in the fight against those things coming against us – and He will prevail. It is He who is the Captain of Battle, not us, for Zechariah 4:6 says, "*...'Not by might nor by power, but by My Spirit,' says the LORD of hosts.*"

The day all this happened came and went, and I did not see anything spiritual in it at all. It was just an unusual event to me for a long time. Then years later while mediating with God, this event came fresh into my mind with the accompanying spiritual points just mentioned. The Lord God gave me one of His significant impressions about that day; the snapshot of that event was taken many years before, for developing at a later time. In the later years I would come to call those significant impressions '*God*Prints*' and write about them in a book titled with that exact name.

As we travel this road of life, it is imperative that we have a Captain of our Soul onboard with us, and His name is Christ Jesus.

Spectre Gunner

At the time of these following incidents, I was a U.S. Air Force Aerial Gunner in the mid 1980s with the 711th Special Operations Squadron (Spectre Gunships). Special Ops Gunners are a strange breed of aviators. They get bounced all over the aircraft cabin and the sky, and love it. The smell of gun smoke mixed with avgas tainting the air is an exhilaration heading into any aerial gunfight.

With one aerial incident, it was a mid-August day; our gunship was participating in a tactical sortie, and I was operating the Number One Gun, a forward 20mm Vulcan Gatling Gun. After about five minutes into engaging live-fire with the acquired target, fireworks suddenly went flying from my gun as pieces of shrapnel went everywhere. This particular Vulcan Gatling Gun fires 3,000 to 3,500 rounds per minute with a loud zipping purr. So, when fireworks appeared below the gun mount, and with accompanying popping sounds – yeah, something definitely misfired.

The Lead Gunner just happened to be right behind me when the fireworks lit off. Sergeant Weldon instinctively grabbed my shoulder, pulled me away and sharply down to the aircraft deck (floor) into the sea of spent 20mm shell casings. While doing this, he screamed 'Cease Fire' into his headset mike. The routine procedure for gun emergencies was calling for a Cease Fire and immediate safety of all guns.

After the prescribed gun cool-down time and an emergency landing on a deserted woodland airstrip, the aircrew observed one of the six barrels of the Number One Gun had 'exploded.' Later investigations revealed that the rifling of one gun barrel had come apart and lodged a spent 20mm round. The next spent round impacted the lodged round stopping the succeeding rounds. The gun barrel literally exploded like a grenade and damaged the adjacent gun barrels. During that flight, gun barrel shrapnel was propelled over the gun mount area, the gunner's station, and near some flight control hydraulic lines. Fortunately, no serious damage to the aircraft structure, flight control equipment, or the flight crew had occurred. Miraculously, the exploding gun barrel shrapnel did not find me.

Yes, I know there are many stories of near misses of all kinds in various avenues of life (and I even have more myself). Some people give God the credit for sparing them, and some do not. It is my choice in giving God the credit for sparing not only me personally that day, but also the lives of my fellow crewmembers and the potential ill fate of that Spectre Gunship. Once upon a time, I was a Navy Aircraft Examiner; and thus, my eyes were trained to detect structural defects both in the making, and after the fact, which most eyes would not see. The incident of that day was a very close call, and one given to me as an imprint just as anyone would receive in similar circumstances. For me, though, it became one of God's personal imprints; for it imprinted into my heart and mind some things God wanted me to remember.

One of those God imprints was just as those bullets come propelling in rapid-fire at thousands of rounds per minute, so do we, sometimes talk in 'rapid fire' imposing our will on those around us. We seemingly must have our way, and usually the first step in obtaining it is a spray of overpowering convincing orations. In that gunship when flying tactical night missions and using tracer bullets, the sustained length of rapid-firing rounds would leave what was called a 'cone of fire' lighting up the target area, and a portion of the sky beneath us.

Do we sometimes leave a 'cone of fire' behind us in the aftermath of things we say to others? Does what we say sometimes 'light up the sky' around those to whom we have aimed our words? We need to be very careful of the things we say to others, for once those words are ejected, they WILL never be returned. They are spent forever at the mark we aimed them toward. Additionally, there may be times when a spiritual jam may occur and our spirit may not be quite up to the level of where it should be with God. Sometimes losing control is easy and then followed by going lighting in with an array of cutting words. During those times engaged in the rapid-fire mode, possible jamming and failure may occur. Then some of our own damaging words can explode right in our faces.

One of God's personal impressions that day was imprinted into my heart and mind; it was that God wanted me to always recall His protection. Regardless of where we are, or what we are doing, we can be taken out of this life in just seconds – here one instant, and gone the next. Yet, God is watching over His children, and He directs the movements of the few for the benefit of the many.

Another incident that happened, which is really something that happens frequently to the Gunner of the Number Six Gun on the AC-130 Spectre Gunships, is being too close to the open cargo ramp. The Number Six Gun is a 40mm Bofors Cannon and is mounted very near the aft cargo ramp of the AC-130 Gunship. Spectre Gunners do not wear parachutes as they constantly get in the way of their aerial tasks. However, at their Gunner Stations somewhere around their machine guns and cannons their parachutes will be located. Those parachutes are strategically placed for last minute grasping and clipping on their parachute harness in the event of egress from the aircraft.

In combat or tactical maneuvers, the AC-130 Gunship flies with the aft cargo ramp in the fully lowered position for various reasons. One reason is the Illuminator Operator (IO) positions himself half way out of the aircraft (tethered, of course) and scouts for incoming anti-aircraft fire and missiles. Another reason for the fully lowered ramp is for fast egress of the crew if the aircraft goes down. Still another reason is for quick ejection of jammed hot ammo rounds.

A jammed 40mm or 105mm round can cook off into an explosion in a matter of seconds. The Spectre Gunner only has a safe window of seconds for clearing the ammo round from the gun, and handing if off to the nervously waiting IO. Then the IO makes a mad dash for the end of the cargo ramp and dumps the ticking time bomb overboard. All the while the entire flight crew is praying to God the fragmentation bomb does not go off and blow a huge hole in the aircraft tail section.

On one tactical gunnery mission, I was operating the Number Six Gun. We were over a sandy terrain and had neutralized the targets for the mission. On the return flight to base, we were given orders to engage another target. Our 20mm ammo was low, so we had to engage the target with a low strafing run of the 20mm guns, then pop up, swing back around, and hit'em with the 40mm cannons. This works real good if there are no anti-aircraft or missile high threats, which in this case our intell information stated there was not.

We engaged the target as mentioned and neutralized it flawlessly. However, the last round in the last clip of 40mm ammo we had planned to fire jammed! I immediately placed the gun in the safe mode and went to work on round extraction. Another gunner came bringing a 105mm casing. The 40mm rounds can be placed inside the 105mm expended casings. When possible, the crew I flew with packed a few of the 105mm casings for just such events as this.

A few seconds later the 40mm round was extracted from the gun, was placed in the 105mm casing, and I handed it off to the IO. Without hesitation, he headed for the cargo ramp's edge, and tossed the 'hot' package overboard. All this time the pilot had been advised by the IO of the round extraction progress, and had been flying toward an area of 'suspicion.' Just co-incidentally, the tossed out round cooked off about the time of impacting the top of a building, and it went up like the fourth of July. Apparently, intell had it that the building was an ammo storage facility. We were to take it out on the return flight, but the when the jammed round cooked off and exploded, that got it instead!

When thinking about these incidents, there are two somewhat applicable passages of Bible Scriptures for life principles, which sometimes come to mind: *"Do not be rash with your mouth...let your words be few...a fool's voice is known by his many words"* (Ecclesiastes 5:2&3), and *"The name of the LORD is a strong tower; the righteous run to it and are safe"* (Proverbs 18:10).

AC-130 Spectre Gunners in flight gear and manning the aft aircraft cabin cannon.

A U.S. Air Force AC-130 Spectre Gunship retired from active service and on static display at the U.S. Air Force Museum in Dayton, Ohio. The author flew as an Aerial Gunner on this same gunship.

'What was'at?!?!?! Anybody hear an explosion through all the gunfire? We hit? We goin' down?'

Rescue Sniper

Once while assigned to a military unit within the covert layers of Special Operations, I encountered an interesting assignment. As part of a tactical clandestine intelligence mission, a fellow Special Operations Sergeant and I found ourselves on rocky-sand covered terrain, undercover of night, evading the enemy. We had obtained our objective and were vectoring toward a safe zone for an aerial extraction at 2300 hours. Successfully accomplishing our mission to this point, we had evaded unfriendly locals, both human and animal types. As we were scathing past the outer perimeter of a small village, we spotted a patrol of insurgent soldiers on the other side of the village about a klick from it.

Crawling over the rocks and across the sand as fast as possible, in as low profile as possible, we headed toward the tree line looming about two klicks before us. It was not long afterwards that the enemy patrol was coming close to us and their presence greatly slowed our progression in the rocks and sand. Luckily, however, was the increase of brush and other tropical foliage along the way toward the tree line. Unluckily, though, at one point the enemy patrol came to stop with a few feet of us.

The soldiers disembarked the trucks and began entering the brush, searching for us. As luck descended on us, we both unexpectedly rolled into a dry ditch about four feet deep, which surprisingly led toward the tree line. The ditch offered us cover as we could move a little faster once passing further from the area of where the enemy had stopped. Once reaching the tree line, we could see the patrol was still searching the area behind us. Going into the trees provided better cover for the next couple of hours, until the time for the next contact call. Our safe zone for extraction was just beyond the other side of the trees, but the area was not so 'safe' anymore.

The time came for contact, and it was determined the helo evac would be made, but in the middle of the wooded area in a very small clearing, and at 2330 hours. If we could not be there, we were to confirm and go stealth for twenty-four. We made it to the clearing at 2330 hours, only a couple of moments prior to the whopping of rotor blades and the dark shadows of two Night Hawk Helos filling in at the treetops. However, a few of the enemy patrol arrived near our safe zone also, for just after the Para Jumper (PJ) dropped down from one Night Hawk with the Rescue Penetrator, those insurgents opened small arms fire on us.

The other Night Hawk returned fire on the insurgents as the PJ grabbed my Sergeant buddy (he was the closest to the PJ) and immediately hoisted him skyward. Seconds later the Rescue Penetrator was descending, but without the PJ. He was leaning from out the helo rescue door with a rifle, motioning for me to grab the penetrator quickly for the helo had to depart, and now! Reaching for the penetrator, I quickly wrapped myself around it and hurriedly begun strapping in. Just then, I felt two hands firmly grab my left boot. Looking down, an enemy soldier had latched onto me and had begun violent attempts at pulling me from the penetrator.

Some might reason it odd that the enemy soldier did not shoot me during the extraction, but took serious risks for a capture. However, consider that what a combatant behind lines has gained knowledge of, and what that combatant knows of their own forces, is of great value to the opposing forces. Taking a combatant alive in some scenarios can sometimes be rewarding for the opposing forces. Kicking at the soldier with frenzy, his grip remained locked onto me as I felt myself slipping from being poorly attached to the penetrator. Looking up, I saw the PJ draw rifle aim on the soldier and take him out.

The next thing to happen was me being jerked skyward as my body flipped upside down with the penetrator! My hurriedly poor attachment to the penetrator, along with too much cable being played out, had ended up flipping the penetrator once it was airborne, and with me dangling from it! Then I met several treetops while being dragged through them, as the Night Hawk rapidly peeled away from the 'once thought' safe zone. Within a few seconds the rescue hoist had me level with the rescue door and the PJ was maneuvering me inside. Placing me on a jump-set and connecting a comcord to my helmet, the PJ's voice came crackling through, 'Hey, man, that was close! But we gotcha all. Stay strapped in and silent. We ain't outta the clear just yet!' And with a slap on both our helmets, he rose to man his machine gun at the rescue door.

Weeks and months went by before thoughts came to me of how similar God and His angels are with this true-life scenario. We go through life evading evil as best we can. Evil, though, is always hot on our trail wherever we go. It makes no difference whether we are Christian or non-Christian – satan does not discriminate. He hates all of us. He just deceives the non-Christian longer than the Christian. There can come times when victory over some burdensome aspects in life may be right before us. It may even seem we are about to be lifted up and away from the troubles we have so long encountered. Then, from out of nowhere, the hand of evil firmly plants its grasp on us, attempting to take us down from victory at all costs.

It is then, in the face of total defeat, that God's power shines forth. For when all efforts have failed and been lost, it is His power and deliverance that outshines them all. God can dispatch a Heavenly Sniper to take out the hand of the enemy, removing it from off our heels, and freedom is gained at last. God has His Heavenly PJs, and they are forever pulling Heavenly Sniper Duty around us. We just never see the results, or the collateral damage, most of the time.

That aerial rescue was not only an actually rescue, but is also contained an impression from God that day, but I did not see it then. Its spiritual message was seen months later. Its full impacting imprint was not seen until a few years later, when evil would come literally against me in the physical, and in the spiritual realm. It would be in those times I would have to call on the Name of the Lord, and He would send His Spirit to dispel the dark evils of satan.

"Are they not all ministering spirits [God's angels] sent forth to minister for those who will inherit salvation?" (Hebrews 1:14).

"But the Lord is faithful, who will establish you and guard you from the evil one" (II Thessalonians 3:3).

The author passing the front of an Iraqi tank

Reconstruction Duty

Ask any military individual who has seen the gruesome devastation caused from war (or natural disasters), and they will verify the need of detailed planning for reconstruction efforts. In any post war reconstruction, if planning is accomplished thoughtfully well, its meticulous efforts will build solid partnerships between countries for many decades into the future. One of many U.S. Army soldiers saw the results of those foundational plans being solidly laid in place for the Philippine Islands after WWII.

Finley Thompson of Louisa, Kentucky enlisted in the service of his country and departed for the U.S. Army on 12 February 1945. Later that spring, he went through Infantry Training in Texas where the trained equipped him as an Infantry Rifleman 'BAR Man' before shipping off overseas. A BAR Man was a soldier who carried a large Browning Automatic Rifle (BAR) in the platoon or squad of GIs. The BAR was invented by gun inventor John Browning in 1917 and adopted for use by the U.S. military in 1918, thus its designation becoming the M1918 BAR.

By 30 June 1938 modifications were made and the U.S. Army designated the rifle as the M1918A2 BAR. The M1918A2 BAR was a longer range weapon than the typical .45 and M-1 automatic rifles. Being of larger caliber (.30-06 Springfield cartridge), the BAR could also inflict more damage. At the beginning of WWII, most U.S. Army Infantry companies designated two or three of its GIs as BAR teams. One man was the machine gunner and one or two men were assistant machine gunners (ammo bearers) who carried extra loaded magazines for the gun. By 1944, some infantry units were using one-man BAR teams, with the other riflemen in the squad detailed to carry additional magazines and bandoliers of .30 caliber ammunition.

Finley was with several other army infantry soldiers on bivouac in a Texas Army field on 12 April 1945. After being on maneuvers for a few days, eating C-rations, sleeping on the ground in tents, and shooting their BARs, their training instructor permitted them a much needed longer break before making evening camp. Finley and a couple of his buddies, nearly worn out, were laid back against their packs and soaking in the cooling southern breeze. About sundown Finley and his buddies began noticing some dust rising on the horizon. As they positioned themselves for a better look, all of the BAR trainees began seeing the rising dust. The rising dust coming toward them was characteristic of the trailing dust from a rapidly advancing vehicle of some sort. Within a few moments, a Staff Jeep arrived in their camp and from it jumped their Commanding Officer (CO). Their Platoon Lieutenant met the CO, exchanged proper military courtesies, and then ordered the men to gather around their CO.

As they all gathered around the CO, it was plain to see from the expression emanating from his face, that bad news was about to be given; and it was. The CO related to the men that he was making rounds to his men in the field that President Franklin D. Roosevelt had passed away earlier in the day. Such news struck the men with shock. President Roosevelt had given the USA his New Deal bringing it from the ashes of The Great Depression, and an appreciative nation could never forget the blessings it brought. President Roosevelt had united and bonded the entire USA together during a time of World War into nearing a time of winning that war. And now, the hero of the greatest country on the face of the present earth had vanished from its most important pages of timing.

As the CO drove off to another gunnery detachment, Finley's unit began silently pitching their tents for the night. It was a most solemn tent pitching for crying was even seen among some of the men. Some of the men who were seen wiping tears were ranged in ages from teenage years up into their forties. The beloved Commander in Chief of the USA had fallen. Such a tremendously unspeakable loss! As the days dredged on, the remainder of BAR training passed even more somberly. The course for the USA was unsure for several days as its leaders continued forging the path President Roosevelt had set in motion long before.

Later that summer, Finley found himself at Fort Ord, California waiting to be deployed to the Philippine Islands. One evening the Officers Club hosted an event for all Officers and men about to be deployed. The event was a performance put on by some Hollywood celebrities, of which Shirley Temple was one of them. Finley was able to meet Ms. Temple as did many of his fellow soldiers after the evening's performance. Whereas the death of President Roosevelt generated some amount of uncertainly toward the future, this meeting with the vibrant Shirley Temple created a catalyst toward healing unity again.

The USA, the news media came alive one day with a most welcome announcement for the entire world. The Empire of Japan announced their surrender which effectively ending WWII. The initial announcement of Japan's surrender was made in the afternoon of 15 August 1945 from Japan. Because of time zone differences, the announcement was 14 August in the United States, Western Europe, the Americas, Australia, New Zealand, and the Pacific Islands. Japan's announcement of surrender created an even greater catalyst toward healing unity again. With that announcement of surrender, duty assignment planning in the Philippines began changing for the GIs headed for deployment there. Finley was given orders assigning him to the 50th Military Police Battalion in Manila. On 21 September 1945, Finley departed the USA from San

Francisco, California, aboard a reactivated U.S. Merchant Marine ship.

A few days later, Finley encountered his first taste with 'sweating out' a very close brush with death. Even though Japan signed the documents of formal surrender on 2 September, it would be weeks before the 'official word' was related to all the Japanese troops in the field. Therefore, all military personnel in the field still had to remain alert and cautious for many months. Finley's ship was crossing the International Date Line in the Pacific Ocean, it was dinner time, and most of the ship's crew and accompaniment were being served their chow (dinner). Finley was in the galley below decks and standing in a chow line when General Quarters (GQ) was sounded.

During GQ all ships accompaniment personnel were to muster on the topside deck with their combat gear and await further orders. He and they all did, and remained topside all evening, night, and into the next morning. What had happened was the ship's sonar had detected a magnetic mine afloat in the area. Therefore, the ship had to stop engines, float adrift, and all the ship's available crew searched and scanned for the mine so it could be destroyed. The mine was finally spotted about noon the next day after being detected. Anti-aircraft guns opened up fire on it, and it exploded like the fourth of July fireworks off the starboard bow (forward right side of the ship). Lunch that day was not a well prepared one, but needless to say, lunch that day tasted exceptionally good!

A few days later on 9 October, Finley's ship docked at the Port of Manila in the Philippine Islands. Prior to docking, they all were informed that alongside their ship, another ship would be docking which would be transporting General Douglas MacArthur. The General (as General MacArthur was commonly addressed) had just recently presided over the official surrender of the Japanese Empire to the Allied Powers on 2 September on the Battleship USS *Missouri* in Tokyo Bay. The General was returning to Manila for only a few days for an in-country engagement and local post war reconstruction planning.

When finally disembarking the Merchant Marine ship and walking down its gang walk, everyone was straining their eyes toward the direction of the ship docked beside them. Everyone wanted that once-in-a-lifetime chance for seeing General MacArthur. About half way down the gang walk, Finley spotted General MacArthur. The general was unmistakable. No one else walked with that gate, had that smoking pipe, wore that style of sun glasses, or cocked their head like General MacArthur did. Those were his unique signatures of character. Everyone had been looking for a crowd of people around an officer. However, the general had come in quietly this time. There were no celebrations and no Manila officials greeting him; only the general and his aides were quietly walking to a staff car. It was this rare scene of quiet confidence surrounding General Douglas MacArthur that stuck more assurance in Finley about the new direction the USA was heading after the sudden death of its beloved Commander in Chief.

Weeks later, Finley had settled into his new routine the U.S. Army had assigned him to in Manila; that of being a Military Policeman (MP). Finley's specific assignment was being paired with another soldier as members of a group of 'Highway Patrolmen' for Manila. He and Corporal Parker would drive the roads of Manila and enforce Military Law on what was left of its roads and country sides. The back roads and the hills of Manila was where Finley saw firsthand the effects war had on people wanting to be free. He never forgot the things he saw – not ever.

When Finley enlisted in the army, most everyone knew the war in the Pacific was winding down. By the time he was being shipped to the Philippines, all soldiers going over knew they were replacements for the war weary GIs coming back home. Soldiers who had fought bitter battles were being replaced by some soldiers who would encounter the bitter aftermath of reconstruction tasks. The hell of war is a hideous ghost, and is NOT complacent before, during, or after its presence.

Finley observed bombed-out, towns, buildings, streets, roads, and bullet riddled everything. He saw fire scorched earth, vehicles torn asunder and piled everywhere, and smoke rising from somewhere most every day for his first several days in country. Bomb craters pockmarked the terrain as if they were common place, and in many places trees were standing shredded into pieces. Everywhere Finley and Parker made their rounds in performance of their highway patrol duties that country's scenery reeked of being ravaged by war. For months overbearing one thought captivated his mind – *destruction on catastrophic proportions.* War torn Manila had been left nearly a pile of rubble. In WWII, only Warsaw, Poland experienced greater damage than Manila. In Manila, seventy percent of its utilities, seventy-five percent of its factories, eighty percent of the southern residential district, and the entire business district had been destroyed.

Reconstruction MP Duty was not really so bad, though. It was just that for weeks after the war ended, the Philippine hills and mountains still contained Japanese soldiers who had not received any official communication; that of the surrender of the Japanese Empire, the cease fire, and the war ending. Thus, attempts at maintaining peace and civil order came with many challenges when renegade Japanese soldiers had to be hunted down, arrested, and transported for confinement in the Allied Prisoner of War (POW) Compound in Manila.

Four years of intense bitter warfare on two world fronts had taught American soldiers grave lessons about enemy appearances – where the enemy was concerned, appearances could be ***fatally*** deceptive. As Allied and Philippine patrols pushed out into its isolated areas, pistols and carbines were ready and faces were always grim, despite the fact that duty called for rounding up the enemy stragglers and dissidents from the hills and mountains in the name of peace, civil order, and reconstruction. Most of them, when learning the war was over, surrendered. However, there were always a few hardcore dissident rebels refusing surrender. It was those hardened dug-in combatants who met their final end in the hills, and it was usually with bullet holes.

Through all the rousting out of Japanese rebels, guarding the roads, and enforcing civil and military law, Finley recalled an odd event with a sailor in a roadside tavern one night. He and Corporal Parker had emptied their canteens of coffee and, still wanting more to drink, were looking for coconuts by the road. After a while of not finding any, pulled their Jeep over at a roadside tavern, walked in, and sat down at the bar.

They both ordered their canteens be filled with coffee, and then a cup of coffee each. About the time they were finishing their cups of coffee, a U.S. Navy man got up and quickly departed the bar. Seconds later they heard their Jeep start up and drive off. After slamming payment down on the bar counter – out the bar doors they ran. To their sad surprise, the sailor which had just exited the bar was making off with their MP Jeep down the road!

Luckily, a soldier who had seen all this had an army truck and offered pursuing the offender with the MPs. And so, the soldier and the MPs jumped into the volunteered army truck and a 'hot pursuit' of the vehicle thief was commenced. But alas, with all bends, twists, and turns in the road (not to mention all the bomb craters in it also), the truck could not catch up with the fleeing Jeep. However, too much alcohol, cratered roads, and blown out bridges never mix well for facilitating fast get-a-ways. A little further down the road the mysterious Jeep-nabbing scenario came to an abrupt ending.

There was a bomb crater a few feet before the remains of a blown out bridge.

In darkness of the night, the bomb crater had swallowed up the Jeep, but without any occupants. The sailor had been thrown out and over the other side of the bomb crater. The Jeep was wrecked beyond repair, but the sailor man did appear salvageable – though he was unconscious. Even though an arrest and a report had to be made on the incident, it was one of the more pleasant ones compared to those involving dissident Japanese soldiers.

Even though the highway patrol duty had its disadvantages when it came to rousting out enemy stragglers, the people of Manila made up for those inconveniences. The people of Manila were so very kind to the Allied soldiers, especially the U.S. soldiers. They always went out of their way to assist the Allies in any way they possibly could. They were finally free from the gripping oppression of the Empire of Japan, and the Allied Powers helped them attain that blessed freedom. General MacArthur may have been forced to leave their country in the beginning of WWII, but he said he would return. The General kept his promise and he did, and with thousands of GIs from the USA with him. Those GI foot prints of the red, white, and blue were gladly welcomed all over the Philippine homeland. It was those foot prints of war weary GIs that set them free.

Summer 1946 came and Finley received orders for going home and for being discharged from the U.S. Army. While standing in the process line to board a ship and leave Manila, Finley broke out with Malaria fever. Subsequently, he was hospitalized, missed shipping out for home, and was transferred to another hospital in Baguilo City for recuperation. Finley was very concerned because he wanted to be home for Thanksgiving and Christmas that year, especially for

Christmas. Having come down with Malaria, his fate was unsure for no one can be of any certainty with the course the illness can take.

Interestingly about Baguilo City, is that it has an altitude of approximately 5100 feet above sea level, and is in the Luzon Tropical Pine Forest. An interesting WWII fact about the city, is that on 26 April 1945, troops of the 2nd, 11th, 12th, 13th, 15th, and 16th Infantry Divisions of the Philippine Commonwealth Army, the USAFIP-NL 66th Infantry Regiment, and troops from the US Army 33rd and 37th Infantry Divisions entered the city. These combined forces fought against the Japanese Imperial Army forces led by General Tomoyuki Yamashita and started the Battle for the Liberation of Baguio City. Later in September, Baguio City became the site of the formal surrender of General Tomoyuki Yamashita and Vice Admiral Okochi, giving up the entire Imperial Japanese Armed Forces to American authorities.

After spending two weeks in a Baguilo City hospital, Finley recovered from Malaria well enough for receiving orders for shipping out back to the States. On 30 August 1946, he departed the Port of Manila, the Philippines, on an army transport ship, and landed at the Port of San Francisco, California, on 15 September. From there he was sent to Fort Sheridan, Illinois, and continued MP duties until receiving an Honorable Discharge from the U.S. Army on 18 November 1946.

Private First Class Finley Thompson was my mother's oldest brother. My Uncle Finley was one of the kindest men I've ever known. He came home from the Post WWII Philippines with a view of war most individuals never attain. That view is people *need reconstruction* from their war-torn, *decimated lives*.

My Uncle Finley, for as long as I've known him, has been a Christian and has followed Godly principals of living. Because he became an eye witness to wide spread devastation, he wanted to continue impacting humanity with reconstructive efforts beyond the war.

Thus, he became diligently involved in Gospel singing groups and working with food banks. Even at the time of scripting this book, he is approaching his eighty-sixth birthday and still works in church and community activities striving toward helping other people better their lives. One of his cherished activities for many years has been direct involvement with local food banks. It still is, and he is still very active in that interest. For years, Finley has willing donated much of his time and resources for assisting others in helping themselves improve their lives.

In the aftermath of WWII, Finley Thompson eye witnessed and experienced people needing personal reconstruction and God. He knew the one and only Book of Specifications which would keep his life on track – the Holy Bible of God. It was taught to him by his parents and, after what he had seen in the post WWII area of the Philippines, remaining within the Bible's precepts was his decision for life. To date, for approximately sixty-six years he has been a *'living reconstruction'* for his God and his fellow man.

To have a life which has been lived as a *'Living Reconstruction'* for his God and their fellow man is a rare epitaph at the closing of God's Books which He keeps on each of us. How many of us can claim such an example about our lives for God? We will all claim something at life's end; what will those claims be? The choice will be ours, and ours alone.

I am the master of my fate: I am the captain of my soul (from *Invictcus;* by William Ernest Henley).

"Do not lay up for yourselves treasures on earth, where moth and rust destroy and where thieves break in and steal; but lay up for yourselves treasures in heaven, where neither moth nor rust destroys and where thieves do not break in and steal. For where your treasure is, there your heart will be also" (Matthew 6:19-21).

Finley Thompson

Photo was taken before he departed for the Philippines in 1945. It is the better of only two known pictures with him in military uniform.

Anzio Road

In March of 1943, Roy Coleman was nineteen years old and living with his parents in Wilcoe, West Virginia. Being very ambitious for trying to get ahead in life, he was working for the railroad and also working part time with his father at the Wilcoe Post Office. Ambition, properly directed, will usually provide the needed gateways for launching bright futures. Other times, ambition will provide an early warning of detours for planned expectations. And so it was, at the Post Office Roy received notice of life's detour; his Draft Notice for the U.S. Army.

Roy was raised in a Christian home, and attended a Church of God with his parents. In his earlier teenage days, he had tried being a Christian but, for some reason, Christian living was not a priority at that phase in his life. Like so many other youth, he had plans for a life destiny awaiting him, and he had to find it. Roy's parents were concerned to outright worried, that their son was going off into the war, and his heart was not where it should be with God.

Roy left for the army departing from Welch, West Virginia, by train and went to Fort Thomas, Kentucky. From there he was transferred to Camp Blanding, Florida, and then to Camp Joseph T. Robinson in Arkansas. After completion of army boot camp, Roy was trained as an Infantry Rifleman. It was while he was in infantry training at Camp Robinson that he learned of being placed on the Overseas List. After completion of infantry training, Roy was transferred to Fort Mead, Maryland. There he was assigned to the U.S. Army's 45th Infantry Division and was also issued overseas supply items. The first part of June, Roy found himself in Newport News, Virginia, being readied for immediate deployment to North Africa.

By the time most GIs were in the Port of Newport News their second day, they were given their orders in sealed envelopes, which were most always carried in their money belts. Even though the GIs who were assigned to disembark on an Allied troop transport ship, they were never informed of their actual overseas port of call. The local rumors had it, though, that their final docking point was Casablanca, in western Morocco of North Africa. When the day of departure finally arrived, it was a bright, sunny, dry day toward the end of September. That day over four thousand GIs boarded the transport ship and it sailed from the shipyard docks. Its only escort was a dirigible; a flying balloon, or blimp. The blimp's responsibility was scouting for enemy ships and then radioing for Allied assistance. The blimp only sailed part of the trip and then turned back to the USA. At the point return, a destroyer escort sailed with them until they landed in North Africa.

Since the transport ship was sailing without convoy, its top speed would hopefully outrun German submarines. Of the nearly five thousand souls on board the ship (including ship's crew), only the ship's crew were permitted above deck the entire voyage during the day. All troops were restricted below deck during the day, and black out conditions were enforced during the night – not even a cigarette was permitted from dusk till dawn. Other than the destroyer escort, not one other ship was sighted the entire eight-day voyage. Luckily, their South Atlantic voyage was uneventful for both weather and enemy activity.

The transport ship came to dock at the Port of Casablanca, Morocco, the very last of September 1943. After days in the sea and the fluid feel of a transit foundation, the sights of houses, buildings, roads, and the feel of solid ground beneath one's feet were a most welcome sensation, indeed! This was Africa, and not America, but at least it was land! After off loading, the GIs were taken to different housing arrangements.

Some were escorted to exhibition halls and government buildings, but all were taken into public facilities – and, the facilities were all surrounded by barbed wire! Some of the GIs joked about if the wire was to keep the locals out, or keep the GIs in, and they would all laugh about it.

After a couple of days in their 'temporary housing,' most of the GIs were taken to the Casablanca Train Depot and loaded onto 'cattle-cars' of an old WWI train. The train load of GIs and their equipment were transported by rail over the North African desert to specific locations. Roy's unit was transported to a location near Algiers. The train trip lasted about three days, and part of the time it was through rain. The train made stops along the way for fuel and rations. Continual smoke from the train engine infiltrating the train cars caused sickness with several of the GIs. Once arriving at their destination, they awaited passage to Italy. After a few days lapsed, the GIs boarded an old P&O Liner and were sailing across the Mediterranean Sea for the Port of Naples.

After landing in Naples, Roy's unit trained for about a month for the mission which was before them. Part of their training in Naples took place in an old racehorse track. The GIs would set up camp inside the track and bivouac from it out to other places. Their training was in preparation for their coming overall mission; that of moving up the coastline and into Northern France and then on into Germany – driving out the Nazis everywhere they went. Every place that Roy trod, the Italian people were most appreciative of the American GIs. The Italians were glad to be freed from Nazi oppression, but they all shared the same main worry – retreating Germans might kill both them and their families. That main concern could be seen soaked into the eyes of most every Italian the GIs met.

Before Roy and the others joined the main body of the 45th Infantry Division, from 10 September to 15 October 1943 the division had secured the area from

Salerno through Alife. The 45th had assaulted Salerno and engaged the Germans in heavy fighting at Ponte Sele; seized the heights overlooking Eboli; pressed along Highway 91 into the mountains, and then took the town of Olive. The 45th pushed onto the Calore River and took over the Bridgehead despite heavy rains and German devastation; mopped up conflicts at Titerno Creek, and connected with the 34th Infantry Division across the Volturno River. The 45th Infantry Division was then withdrawn into reserve on 20 October.

After the month of additional training Roy and the others completed, they were integrated with the main body of the 45th Infantry Division. The division had become engaged with seizing some very strategic mountain positions from the Germans; those actions began on 29 November. About a week later on 6 December, Hill 769 fell to the 179th Infantry and then Mount La Posta gave way to the 180th Infantry on 17 December. The 180th Infantry then attempted securing the hills north of Cassino. However, through all the fighting during the holidays, and with all the advances having been made thus far, on 9 January 1944 the division was withdrawn for further planning.

During some of those engagements, Roy's unit was bivouacked in an old crater. His platoon sergeant had tasked Roy and two other GIs with escorting close to a tank for taking out some enemy positions. During one tank escort duty, German mortar rounds were fired at them, and before the tank could eliminate the enemy position, one mortar round landed next to Roy. The round did not explode; it was a dud. Later, up another hill, the Nazis were heavily strafing Roy and some others, and Roy was hit in the shoulder. He was taken into the home of a local Italian for first aid. After getting patched up, he was then taken to the area field hospital for more thorough medical treatment. After the battle lulled, the Nazis infiltrated the area, started raiding the hospital, and began killing the wounded before some American GIs entered and drove them out.

As always, plans are conceived, strategies are developed, and then decisive orders arrive. The division moved out again and assaulted Anzio, Italy on the early morning hours of 22 January 1944. The 45th Infantry Division met with strong enemy opposition which caused them to become occupied with defending the Anzio beachhead for the next four months. This offensive maneuver actually turned into a defensive position for a few weeks. It was part of the larger Allied plan called Operation Shingle (22 January to 5 June, 1944). This offensive operation was originally conceived by Winston Churchill for implementation during the Italian Campaign of WWII. Operation Shingle was an Allied amphibious landing against Axis forces in the area of Anzio and Nettuno, Italy. The overall strategy was intended to outflank German forces of the Winter Line enabling an assault on Rome. The lengthy conflict at Anzio became known as the Battle of Anzio.

For about four months the GIs of the 45th Infantry Division held, and defended, an area of about ten square miles of Anzio – and that included the Anzio beachhead. Roy found the foxholes in the Anzio beachheads quite uncomfortable; they always seeped with water, no matter where they were dug on the beach. Therefore, the GIs usually always became damp to wet when spending any amount of time in their foxholes.

One night Roy and another GI were on guard duty along the Anzio beachhead. During part of the night enemy shelling kept them company. An Allied ship was out at sea, and for about an hour or so, artillery fire was being exchanged between the ship and the Germans on land. Then about 0500 that morning, a massive explosion occurred out at sea; the Allied ship had suddenly been hit by German artillery fire, and rapidly sank. Roy and the other GI just stood there on the beach at their post, watching the sight, and thinking how helpless they were to the hundreds of men on that fast sinking ship.

On 11 February 1944, the 179th Infantry attempted recovering areas around the Factory, but the effort was in vain. At the same time, the 157th Infantry became isolated west of the Albano Road by German counterattacks. One night during these events, Roy was assigned to be on watch near an old house in Anzio. He was an Assistant Gunner to the BAR Gunner. Both he and the gunner had set up their gunner's position in a foxhole for the night watch. While securing the area near an old house, and returning to their foxhole, Roy did not hear the gunner call out 'Halt' twice. The gunner opened fire. Just as Roy dropped into another foxhole, he felt the passing sizzling slur, followed with a movement of air, and heard the whizzing sound of bullets zip right past the side of his helmet. (Anyone who has ever been in the military in combat, or around gunfire where bullets have came within inches of your head, is well aware of this sensation and sound). Roy was one very appreciative GI to God that day!

On 18 February, the division encountered an intense German assault for regaining the area they had lost. The enemy assault was contained, though, after the 179th Infantry had been pushed back to their beachhead final 'line in the sand'. One day during this German assault, Roy and some other GIs had taken cover in a wet beachhead foxhole. It was so soggy they had to line the bottom of it with their rain coats in attempt to keep the dampness from seeping through onto them. As the battering day wore on, they kept hearing the sounds of incoming artillery rounds plummeting ever closer to them. Then there came a loud whistling sound and it suddenly became silent. When the whistling sound went silent, they all knew the round had found its target, and it was them. What looks that could be shared, were those of ghostly souls waiting for their end to come – just ticking seconds away.

The artillery shell impacted with a penetrating explosion close to their foxhole scattering debris and shrapnel everywhere. When the dirt, dust, and debris settled, the GI next to Roy had been killed as were several others. The calf of Roy's left leg had taken large pieces of shrapnel, and his right calf had been pierced with shrapnel to the bone. Bleeding profusely, Roy called for a medic, as did others nearby who had been wounded. Through the hail of incoming shelling, flying debris, and smoke, medics eventually arrived, patched up the wounded, and got them evacuated to a field hospital. Roy's leg injuries were severe, and by the time Roy received adequate treatment at the field medical unit, he'd lost considerable blood. In spite of sustaining traumatic battlefield wounds, this was another day Roy became a most appreciative GI to God.

Even though the medical staff attempted cleanliness of the impacted area, insidious battleground bacteria had infiltrated an advance foundation into the left leg wound by the time he'd been transported to a British hospital ship in Naples. His length of hospitalization in Italy was always between the hospital ship and an Allied hospital in Naples, but mostly aboard the hospital ship. During his first night aboard the hospital ship, he was in constant pain and continually called for medication to ease it. His first hospital night, though, was spent in anguish for no one attended to him until the next morning. Roy's left leg wound ended up becoming gangrenous, and all sensation below the calf to the foot was lost. The doctors informed him that portion of his left leg must be amputated. The next day after receiving such dreadful information, surgery was performed for removing the affected section of his leg.

Roy's recovery was slow because the gangrene had begun filling his body with infection and draining its resources. When feeling well enough, he wrote one of his friends back home with the information about his leg. Roy requested of the friend to relate the information to his father. This way his father could break the news to his mother before seeing them again. While hospitalized, he discovered that on 22 February British forces came ashore at Anzio and with the American forces, engaged in holding actions against the Nazis in resuming the Allied defensive position of Anzio.

Also while recovering in Naples aboard the British hospital ship, during the latter part of March 1944 Roy spent much of his time on its open ship deck. This was because the ship housed so many wounded soldiers. This open deck actually became his living quarters for several days, and provided him with having a grand view of the rumbling Mount Vesuvius Volcano. Wasn't that a comforting sight? In fact, he saw its historic volcanic eruption from 18 to 29 March 1944. The eruption of Mount Vesuvius that month killed twenty-six people while destroying the villages of San Sebastiano al Vesuvio, Massa di Somma, Ottaviano, and part of San Giorgio a Cremanocaused.

In spite of sustaining traumatic wounds, and having part of his left leg amputated, the days for the eruption of rumbling Mount Vesuvius Volcano were even more days Roy became a most appreciative GI to God. Lava and spewing debris from the exploding mount could have destroyed some of his hospital ship, and either injured or killed him.

When Roy's recovery began improving he was transferred by the British hospital ship to an Allied hospital somewhere off the coast of Morocco in North Africa. While recovering in this hospital, Roy's personnel records (not medical records) were discovered having been misplaced. A staff officer began questioning his length of hospital stay. Subsequent inquires revealed Roy's missing personnel records were in the USA. The staff officer took immediate action for initiating a set of orders having Roy transferred to the USA.

Roy was subsequently transferred to another ship, an American hospital ship, and then on 29 May, Roy found himself leaving Morocco and sailing for the States. The nine-day voyage back home was not as restrictive as the one coming over to fight in the war. This ship was a hospital ship and was unescorted. The only incident was a minor one, and it was at the Straits of Gibraltar. The passage was thought to have been mined and there was a delay until it could be certified clear for passage. It was on the trip home that Roy first learned to walk with crutches. It was also on that voyage home he tasted his first Coke-a-Cola since before leaving the States.

On 6 June 1944 (Normandy D-Day), Roy's ship docked at the Port of Charleston in Charleston, South Carolina, USA. After leaving the port, Roy was taken to the Walter Reed Military Hospital in Bethesda, Maryland. He was sent there for further evaluation of combat wounds, fitting of a prosthetic limb, and rehabilitative activities. When first entering Walter Reed, the medical staff determined the tibia bone of his left leg was splintered too excessively for accepting the artificial limb. Thus, another two inches of his leg had to be removed. Roy was given a spinal tap and the section was removed while he was awake.

Also while at Walter Reed, Roy discovered that when the British forces landed at Anzio after incurring his leg wound in combat, the combined British and American forces resuming their offensive for Operation Shingle led to victory at Anzio on 23 May. The 179th and 180th Infantries then attacked the Nazis further up the Albano Road on 29 May. By 4 June the 45th Division had obtained secure crossings north of Rome, and thus, paved the way for entering Northern France. Their road of victory through Italy and France was still pressing onward to Germany.

Operation Shingle (the Battle of Anzio) resulted in over 40,000 Axis Powers causalities; approximately 5,000 KIA, and 30,500 WIA & MIA. The Allied Powers suffered over 43,000 causalities; approximately 7,000 KIA, and 36,000 WIA & MIA. Private First Class Roy Coleman was one of those causalities, and was Honorably Discharged from the U.S Army after serving one year, eleven months, and five days. He returned home to Wilcoe, West Virginia, in February 1945 and resumed his civilian life. His prosthetic limb received at Walter Reed never agreed with him until the army was able to obtain one from a company in Cincinnati, Ohio. He recalls that fitting episode well, for it enabled him to become more comfortably mobile than before. Counting his days in field hospitals and in Walter Reed, Roy was hospitalized for twelve months before receiving his discharge to go home and resume his civilian life.

As mentioned previously, Roy tried being a Christian in his teenage years, but drifted away from his devotion to Christ for awhile. Even while sustaining assaults, shootings, shellings, and bombings, Roy would pray to God for protection of his life. Back home in West Virginia his parents were also praying for his life and protection. In 1947 a lady minister came to the Wilcoe area preaching a revival and Roy attended one of the services. It wasn't until then that Roy decided it was time he turned his whole heart back toward God and rededicate his life to Christ.

Roy had been writing a lady in the USA while overseas. Once while in the Walter Reed Hospital he was given a week furlough. He came home and also visited the lady. The visit proved the relationship would not work toward anything permanent. Later, after his discharge and then returning home, he met another young lady by the name of Helen (Deanie) Newton. Meeting her sparked an interest so strong, that a match between both of them was inevitable. They were married not long afterwards and stayed married until she passed away in January 1986. That match was, and worked, so well that Roy never remarried.

Roy Coleman became a casualty of war during WWII. The events of WWII changed his life, but he never regretted his service for his country. Roy went where his country needed him and served well. The people of Europe, and especially Italy, can be grateful for his willing sacrifice left his footprints of *the great red, white, and blue* on their lands – and some of those footprints were indelibly etched in their minds for decades to come.

Some hecklers may say where was God when the Nazi artillery shell exploded near Roy and took away the partial use of one leg? Yes, the doubters and the agnostics will always come up with skeptical lines such as that. However, the bottom line about war is: war is war. Bullets *will* fly, bombs *will* drop, knives *will* cut, and swords *will* slice. Why? It is because the warriors *will* such things to be, or either they react in defense of those *willful* actions. Thus, God, as an impartial third party to those *willful* actions of humanity, listens for their *frantic prayers of peace* and guides His angels for the best actions of all concerned. His responsive actions may be for the present time, and/or those of future times.

This is for the heckling skeptics: (1) Wonder *how many people were praying* for the GIs around Roy at the time the Nazi artillery shell exploded? (2) Wonder how *many prayers had been offered* for those GIs around Roy at the time the Nazi artillery shell exploded? (3) What's to say the shell would have been a direct hit in the foxhole had it not been for God's Hand nudging it over? (4) What's to say Roy would have been killed had NOT God permitted the wound for another purpose? That purpose being so Roy would NOT be killed later in the war, or so that he would come home and FIND the love of his life.

The answers for the 'whys' and 'what ifs' in war are incessantly before us. Yet, its questions supply the never-ending fabric for stories of humanity's crusades.

One thing for sure is as previously mentioned, in war bullets *will* fly, bombs *will* drop, knives *will* cut, and swords *will* slice. Why? It is because the warriors *will* such things to be, or either they react in defense of those *willful* actions. Christ Jesus even said there would be wars and rumors of wars (reference Matthew 24:6). Therefore, the obligation falls to each and every person, civilian or military, to be ready for whatever comes with the terrible actions of war. Be ready for acting wisely and leaving this life *ready* to be with the Lord God. Friends and family may pray for us, but the final choices of life are *uniquely* all our own.

"Yea, though I walk through the valley of the shadow of death, I will fear no evil; for You are with me; Your rod and Your staff, they comfort me" (Psalm 23:4).

Roy Coleman
Photo was taken before he departed for North Africa in 1943

Helmet Hit

In 1969, Corporal Chad was in the United States Marine Corps (USMC) serving in Vietnam. His job a few weeks prior to shipping out back to the USA was being a supply escort. This meant wherever a supply shipment went he made sure it arrived. His delivery shipments were USMC guaranteed with an M-16 rifle in one hand, and a shipping manifest in the other hand. Shipments arrived by Jeep, truck, or helicopter.

On this particular day, Chad's delivery was being made by chopper, a USMC CH-46 Sea Knight helicopter; and colloquially known in the Marines as the 'Phrog.' Marines also had another cliché about that chopper; it was 'phrogs phorever' and was because of its dependability in the field. Also on this particular day, after making delivery, they were to double in a medivac capacity. On the return sortie they were to load a WIA, and then fly the WIA to a field hospital before returning to home base.

After landing on some God-forsaken hill somewhere in the An Hoa terrain, Chad oversaw the ground crew off loading his shipping crates from the back ramp of the chopper. Having been informed the landing zone was becoming hot, he signed his own manifests as the WIA was being loaded on board. After slinging his M-16 over his shoulder, he turned to walk up the ramp when they began taking enemy fire. Before the chopper's ramp could close and they could become airborne, Chad heard a ding on the right side of his steel pot (combat helmet) while at the same time felt an impact knocking him into the chopper and onto its deck. Next thing he sensed was something like warm water coming down around the sides of his head.

As the chopper ramp began closing Chad ran his hand across his face thinking to wipe away profuse sweating. To his surprise, his hand was quite coated with blood! Knowing then the ding was a helmet hit, he removed his helmet and to his great surprise, the bottom of the helmet liner was soaked with blood! By this time, nausea mixed with tunnel vision was setting in. As shades of gray intensified, Chad became weak and completely collapsed over on the chopper floor deck. Looking up at the Door Gunner firing away, everything faded from shades of gray into all black.

On the other side of the world on a Kansas farm in the USA, a Christian mother was out feeding her chickens. Chad's mother suddenly felt impressed of God to pray for her son. The impression was so strong, that it was actually *compelling*. It was compelling to the point of her literally dropping to her knees and beginning to immediately pray for her son. After a few moments, she rose from prayer, walked a few steps, but could go no further. The burden for prayer had not been lifted and falling again, this time she fell prostrate to the ground into some hay. She lay there crying and moaning in prayer for her son. After about an hour of intense prayer, the burden was lifted, she arose, dusted of the hay, and went back to her house wondering what calamity had befallen her son in Vietnam.

Back on the other side of the world in a field hospital in Vietnam, Chad awakened. Close to a day passed since hearing the ding in his helmet, seeing it splattered with blood, and seeing the Door Gunner fade from shades of gray into all black. Next thing he realized was having one terrific headache. Raising his right hand to his head, there were bandages all around it alright. But somehow, thank God, he was alive! Laying there in pain and pondering all this, he was visited a few moments later by a Field Doctor coming to his bunk.

'You are one lucky jar head ya know' came from the Doc as he stopped at Chad's bunk.

'What happened Doc, sir?' Chad asked.

Holding up a projectile from an AK-47 round, the doctor began saying, 'See this bullet? It struck the side of your helmet, penetrated it, and somehow entered in between the metal helmet and the helmet liner. Then it traveled spinning around between them until its energy was depleted. As it travelled spinning, it ate up the helmet liner and then began eating up your hair and skin in one single line around your head. That's what caused all the blood in your helmet. You lost a lot of blood and hair, but not too much skin. A few days R&R to let that skin grow back and you'll be a Grade 'A' Jar Head again.'

Chad was transferred to a defense hospital in Da Nang. There he just happened to find the WIA his chopper crew had been transporting the day he took the helmet hit. After about three weeks, Chad was returned to full duty with his unit. Chad had a necklace made from dog tag chain and the bullet hitting his helmet. He wore it with his dog tags until returning home from Vietnam. When returning home, Chad provided a more detailed dialogue with his mother about the strange incident and discovered she had been praying for him. He also discovered her most sincere prayers for him were the *exact same hour* as when he was encountering his helmet hit incident. Chad gave his mother the bullet necklace, and she wore it for a long time as a reminder that God answers prayers.

There will be many who read this story and remark of the thousands of soldiers who did not come home from war – and ask why? All that can really be said is, only God knows why. That short answer is not much comfort, and is not intended as a cop-out; but in reality, it is God who knows only such things far above us. Interesting isn't it? How ***WE*** **make** and wage wars all down through history, and then ***WE*** **fully** expect God to protect all the players in those clashes all the time?

As for this episode with Chad, he became a Full Gospel Minister a few years later. It is my very strong suspicion God had a plan for his life that included him doing God's Work on a Divine Scale. Thus, God intervened when the bullets went flying, especially the one which was 'dead on target.' If nothing of this story is absorbed, and gleaned for a life lesson other than the next two paragraphs, read them carefully, and retain them well.

In other scenarios with other soldiers, how many of them have God-fearing mothers and fathers who are really in tune with the Spirit of God on a daily basis? How many soldiers actually have someone special in their lives back home ***WHO*** will drop to their knees *when* God impresses them that ***NOW*** is the *appointed* time to pray? Maybe, just maybe, soldier safety on the battlefield is more directly linked to devotion on the home front than was ever thought before. Maybe such is only a ***PART*** of the answer as to why some do not return home – maybe, just maybe.

The campfires of the war front may be more *closely associated* to the hearth on the home front than has ever been realized.

Postponed Honors

The 6th of June 1944, is a date etched in the annals of human history, and it became forever known as simply 'D-Day.' That infamous day was when the mighty Allied Armada crossed a thin channel of sea from England to Normandy, France. It was there, on those bloodstained and shelled-out beachheads of Normandy, the Allied Forces of WWII broke the Nazi conquest of Western Europe.

The price some of those Allied soldiers paid for the liberation of Western Europe was a supreme one; one paid with their lives. Other prices paid came with time away from home and loved ones, and some with temporary and/or permanent handicaps. For some, ribbons and metals were given as rewards for the work well done. For others, the world's most important job was never repaid beyond that of knowing what was done was the right thing to do. Some suffered more than others, were never compensated, and were never recognized for their suffering. And, such is the nature of this story.

While the world was focused toward the beaches of Normandy on 6 June 1944, other facets of Allied resistance were in operation for ridding Europe of Fascism and Nazism. One of those facets was the United States Army's 45th Infantry Division; it had been in Italy and was on the move toward France. The 45th Infantry Division had assaulted Anzio, Italy, on 22 January 1944, and was heavily engaged in defending its beachhead for the following four months. On 11 February, its 157th Infantry Regiment became isolated west of the Albano Road by German counterattacks, and one week later the 45th Division came under a concentrated German counterattack. Conflicts continued with the American and British forces fighting the Germans through the end of May when they broke through an area further up the Albano Road. By 4 June, the division secured crossings north of Rome and, on 16 June 1944, the division was withdrawn from combat for rehabilitation.

George W. Philemon of Oakboro, North Carolina, joined the U.S. Army after leaving high school. His first assignment after boot camp and advanced training was the 157th Infantry Regiment within the 45th Infantry Division. He was then transferred to England, and from there he found himself landing in Southern France in September of 1944, to join his unit which had not long since crossed the Rhône and Ain Rivers northeast of Lyon. Blending into the military machine of 45th Infantry Division, George and his comrades assaulted Villersexel, and in spite of intense fighting, the city was eventually seized on 13 September. Then on 21 to 22 September, the 45th crossed the Moselle. The 180th Infantry continued hammering out their marching paths for taking Epinal on 22 to 25 September, while the 157th Infantry seized Girmont on 24 September. Only then was the 45th Division enabled to enter the western foothills of the Vosges Mountains bordering the Black Forest of Germany.

After severe fighting and casualties, Bruyerès fell on 19 October. Afterwards, on 22 to 23 October, with several attempts by the 180th Infantry, a much needed bridgehead was spanned across the Mortagne River near Fremifontaine. On 23 November, the division moved up to the Saverne Gap and captured the forts north of Mutzig in the Maginot Line on 25 November. Moving up to the Moder, the division took Mertzwiller in heavy combat on 5 December. Fighting across the Lauter into Bedenthal on 18 December, the division engaged in battles which isolated several elements; and the division contained the Germans along the Philippsbourg-Neuhoffen-Obserteinbach line through 1 January 1945. The division battled in the Bitchie Pocket and captured Wingen on 7 January, but was forced out of

Althorn by counter attacks on 11 January. On 18 January, a battalion of the 157th Infantry Regiment was decimated northeast of Reipertsweiler.

George Philemon had enlisted in the army to serve his country, and he had, indeed, done just that by fighting across France into Germany. He had found himself serving in the European Theater of War, and later, specifically in the mountains of Germany in what was to become known as the Battle of the Bulge from 16 December 1944 to 25 January 1945. The battle was also known as the Ardennes Offensive; and for the Germans, the battle was known as the Von Rundstedt Offensive. The Battle of the Bulge was so named because of the westward bulging shape of the battleground as viewed on a map. It was the largest land battle of WWII in which the USA directly participated. Records estimated that approximately 500,000 American, 55,000 British, and 600,000 German troops collided in that battle.

At the time, Germany was unmistakably losing the war. The Soviet Red Army was closing in on the Eastern Front, while strategic Allied bombing was wreaking sustained havoc on Germany's cities. The Italian peninsula had been captured and liberated, and the Allied armies were sweeping through France from west to east. German Führer and Nazi Leader Adolph Hitler knew the end was near if something could not be accomplished to slow the Allies' advance. The Battle of the Bulge was his desperate measure for slowing that advance.

In the bitter cold of that 1945 winter invasion force, over one million soldiers clashed in battle along an eighty mile front running from southern Belgium through the Ardennes Forest and down to Ettelbruck in the middle of Luxembourg. George Philemon's 157th Infantry Regiment was approximately eight miles from the Rhine River. George was fighting for his country, trying to stay alive, while at the same time also striving to protect the lives of his buddies. Within those ever growing layers of vicious conflict and nature's elements, those GIs became swallowed up into the intense crucible of the Battle of the Bulge.

In the aftermath of what would become known as one of WWII's most horrendously historic battles, most of George's unit was killed in action. On 9 January, his company was northeast of Reipertsweiler and pinned down in a trench by German machine gun fire, with only 365 GIs still remaining alive. During one enemy attack, George's best friend at the time (Perkins) had tried protecting him from strafing bullets. The shielding attempt proved well for George, but was unsuccessful for his friend. Perkins was mortally wounded when both of them were strafed by machine gun fire and then he died beside George in the trench. His buddy had died covering George's back with his own body. That is a notable sacrifice of the highest order; to give one's own life for a comrade or a friend. All throughout military history its members have commonly addressed each other by last names only. So it was with Philemon and Perkins when they were thrown together in the throes of combat. George never knew Perkin's first name, but he did know that he was married, had children, and was from West Virginia.

The remaining survivors lived in their entrapped trench for seven days. Three days after entrapment, their water supply exhausted and snow was melted for acquiring drinking water. Then, on the fifth day, their food rations were exhausted. The GIs stayed held up in the trench as long as they could. However, no food, no water, and bitter cold can become other pungent enemies as well. Then, after the seventh day, on 16 January the bitter cold won, and surrender to their vigilant German captors became the GI's only option for survival.

George and the remaining survivors were taken as POWs. As any soldier would be, they were humiliated by having to surrender. They were first transported by railcar to a Stalag for processing in as a POW. The German term 'Stalag' stood for their POW Camps designated for enlisted military members. On arrival, the POWs were strip searched, deloused, given POW uniforms, and given a POW postcard with which to write home to a family member. With the end of the war nearing, the POWs were frequently moved and shuffled among the camps. George resided in at least three POW Camps: Stalag 12A (Stammlager XIIA) in Limburg, Germany; Stalag 11B in Bad Fallingbostel, Germany; and Stalag 6 near Heydekrug, Germany. Heydekrug lies south of the Denmark border and between the North and Baltic Seas.

In all of the POW Camps, the German captors required all POWs to work. The usual POW food ration consisted of one bowl of thin soup once a day. There were times food and water would become a little diminished. Therefore, since reduced rations became less than one bowl a day, it was during those times of sparse rations that labor tasks would become very difficult. This was because the POWs became too weakened for labor.

While in one of the POW Camps, George sold a cigarette for one silver dollar. His winter boots had two layers of felt pads in the soles, and he had placed the dollar between the felt pads so no one would steal it. His plans were for bringing the dollar home if he ever made it out of the camp. However, hunger will always become stronger than any desires for retaining souvenirs. So, after carrying the dollar about a month, George used the dollar for buying a rutabaga and feeding the aching hunger always gnawing within his stomach.

George had bathed six weeks before being captured, and did not receive another bath for three months. The German guards would occasionally delouse the POWs. When delousing occurred, the guards required the prisoners to put the same clothes right back on which they had just removed. POW latrines were very crude and simple. They were just two poles laid alongside a ditch. Many of the POWs would laugh at those fragments of simplicity because they'd say 'You don't have to flush them.' Those scraps of occasional humor actually became segments of life-savers. It was there, in those seemingly God-forsaken surroundings, that George made a discovery – finding humor even in bad situations can actually improve one's outlook for just a little longer. Humor may not *find* solutions, but humor can help *relieve* some stress until solutions *appear*. In spite of the cold, poor rations, poor clothing, and occasional bad treatment, humor was a life-saver until being released.

In April 1945, Allied soldiers invaded the last POW Camp where George was detained, and liberated the POWs. The POWs were then transported by trucks to a nearby airport. From there, all of them (George included) were transported to the 91st General Hospital in Oxford, England, for treatment. When arriving at the hospital, the GIs were fed all the pork chops they could eat. Then for the next ten days, their meals consisted of mainly vegetables and boiled eggs three times each day. George was a patient in the 91st General Hospital for about three weeks. In May, he was transferred back to the USA and then, in June, transferred to Fort Bragg, North Carolina. On 26 December 1945, Private George W. Philemon was Honorably Discharged from the U.S. Army.

George may have come back home to the USA from WWII and resumed his civilian life, but removing war experiences from his mind were not as easy as he thought they would be. George's memories of those three months as a POW were things he never talked about very freely. Some things encountered in life are those of suffering endurance, and they are best left alone for a long time. His family never knew of the heroic details of his days in WWII, especially his late wife of 48 years, son Sherron, and daughter Marilyn; they hardly knew anything of his time spent in the military. Those were times George just never wanted to discuss much of its details with anyone. That was his way of dealing with what he went through and putting it all behind him.

Post war life for George became routine until one June day in 2009. In Union County, North Carolina, some veterans are anything but the routine. Over the years, the Veterans Council of Union County has been very active. This group gives much of its resources to veterans in need within their region. They have assisted thousands with food, fuel, transportation, and friendship for many years. Some months before that June day in 2009, a young soldier's name arose in conversations. Over the ensuing dialogs, many questions surfaced as to whether or not he had ever been recognized for his service in the Battle of the Bulge.

Subsequent investigations revealed George had never received a POW award for his time spent in the Stalags. The Veterans Council began working to secure all the medals that George should have been awarded. After months of planning by the Veterans Council, the local veteran organizations, along with the help of the active military, everything was in place for George to receive his belated awards. A special occasion was planned for the 6th of June 2009, at the Euto Baptist Church near George's birthplace at New Salem, North Carolina. Many friends and associates endeavored for making this day a most special occasion.

At his award ceremony, George stated, 'Freedom is not free, a lot of guys died and suffered for our freedom. If we had not done sixty-five years ago what we did, you'd be reading your newspaper in German or Japanese. I am proud to have served my country; even though I did not want to go, I'm glad now that I was a part of history.' George W. Philemon was awarded eight medals by Colonel Scott Kerr and Captain Ryan Stafford. The awards included: Honorable Service Lapel Button WWII, Infantryman Badge First Award, WWII Victory

Medal, European-African-Middle Eastern Campaign Medal with Bronze Star, American Campaign, Good Conduct Medal, Prisoner of War Medal, and Bronze Star Medal. These awards were long overdue for an American soldier and a man of great loyalty, who served his country way beyond his call of duty.

Every American should be thankful to all of its veterans who have served this country in the past and those who are actively serving. Without their willingness to *sacrificially* serve the principles behind *the red, white, and blue*, Americans and much of the world ***would NOT be free***.

On another level of patriotism with country and home fires is that of God and His people. Humanity fails to recognize that God is a patriot; and He, too, has honor and devotion to duty. John 3:16 is the most resounding proof of God's honor and devotion to humanity – "*For God so loved the world that He gave His only begotten Son, that whoever believes in Him should not perish but have everlasting life.*" Just as a soldier in the Battle of the Bulge gave his life for his friend George and his country, God gave the life of His Son, Christ, so that humanity can be restored back to Himself and never be lost. That is patriotism of the highest order, a Divine One.

Another level of Divine Patriotism is that of rewards. God never forgets, or omits, any reward due His people. In this story of George Philemon, some long overdue honor and recognition was rendered him sixty-five years way late. His sacrifice and service was forgotten by his country and by many until a faithful few would not let those things pass on by any longer.

With this world, it may forget about the *seemingly* insignificant things we accomplish as the years fade their values away. Those things we achieved as important may become as dust in the wind within the chronicles of time. However, the things offered for God, completed or failed, will never be forgotten by Him. God keeps an eternal record and will one day give a reward for those things offered for Him, regardless of how small or how great they were (reference Mark 9:41, Mark 12:41-44, and Revelation 20:11-15).

In general, the world scoffs at Christians, their God, and His Son, Jesus Christ. But, one day will be the Universal Reward Day for everyone. On that day, the <u>*appearance*</u> of long overdue postponed honors will be cast aside forever, and those eternally earned honors will be bestowed. For those who have not God and Christ, what will that day bring? Will there be honors, or will there be something else less?

"...Be strong and of good courage; do not be afraid, nor be dismayed, for the LORD your God will be with you wherever you go" (Joshua 1:9).

George W. Philemon
Photo was taken before he
departed for Europe in 1944

Them Two Bullets

The 2nd Marine Reconnaissance Battalion falls under the Second Marine Division and the Second Marine Expeditionary Force (II MEF). 'The 2nd Recon' is what the Marines proudly call their unit. Their motto is 'Swift, Silent, and Deadly' and their combat history proves that motto exceptionally well. The 2nd Recon has stealthily gone into multiple places 'unknown,' accomplished 'peacekeeping missions,' and came home time and time again singing hoorah all the way.

However, it is while back at home base GIs –and especially Marines– are well known for playing heartily. Built up steam from combat must be vented; everyone knows this fact (or should, anyway). And so, the Marines must take turns assisting the base Military Police (MP) with their peacekeeping duties at home.

In the 1970's, a U.S. Marine by the name of Gunnery Sergeant Mick was serving with the 2nd Recon. Gunny Mick was pulling MP Duty with a Staff Sergeant one Monday night. Their shift had gone relatively well for a weekday evening at their base. Routine checks were carried out in seamless order along with putting down a couple of fights for the evening. The fights had broken out in the enlisted club parking lot. When the MPs arrived, the conflicts very soon were squelched by the Gunny's overly friendly manner of persuasive points of views – 'Clear out in five, or I'll hall ya'll's butts to the brig and ya can cool off in there! It's ya'll's 2nd Recon choice gentlemen!' When it came to making decisions, especially crucial ones, the Gunny had no problems at all; and his thundering words always followed them, too.

However, the parking lot fights were not the most interesting event of this particular Monday evening.

About 2230, Gunny Mick and his MP sidekick were toward the end of making their last security round (their shift ended at 2330), when a call came through on their radio. A disturbance had been reported by one of the base barracks Master-at-Arms, and he had requested MP intervention. Within about five minutes, Gunny Mick drove the MP Kaiser Jeep up to within a block of the barracks. Calling back to MP HQ, Gunny Mick got an update on the disturbance. Apparently some Marines were in the rec area in back of the barracks loudly playing a boom box, and loudly singing along to some songs. After receiving the update, Gunny Mick had his sidekick drive around the back side corner of the barracks while he walked into the front entrance. He stood at the rear doorway and observed about seven or eight Marines gathered around a table in the rec area and listening to a boom box.

While Gunny Mick stood in the rear doorway of the barracks, the Marines in the rec area saw the MP Jeep and began whooping away at the MP driver. The MP driver exited the Jeep and just stood beside it as instructed. At this, the Marines began loudly laughing, and those actions began ticking things off for the Gunny. He immediately opened the barracks door and walked straight to the rec table, turned off the boom box, and plainly said, 'Gents, Taps was sounded over thirty minutes ago. It's time to tone things down.'

'Gunny, what's the big deal anyway? We just havin some extended rec time' came back a reply.

'Gunny, why ya bustin up our fun?' came another reply.

'Gunny, ya just can't go switchin' off my boom box like that. It ain't yours' came the boom box owner's reply.

'Gents, am gonna say this one more time in more detail so ya all will understand it. Taps is sounded so everyone can get a good night's rest after a hard day's work in the field and all them other God-forsaken places we are sent to. Ya all can play and party hard from Reveille to Taps, but after Taps ya gotta respect the rest time of others. Taps was sounded over thirty minutes ago. Leave that boom box off or either the volume down low. Is that perfectly clear?' came snapping through the Gunny's teeth.

After a few seconds pause, the boom box owner reached down, switched the boom box back on, and then turned its volume back up to its previous volume. He then looked the Gunny sneeringly in the eyes as if to challengingly say – hey dude, what's ya gonna do now? Without hesitation, and with smooth symphonic motions, Gunny Mick stepped back; unsnapped his pistol hostler; drew his weapon; cocked it; aimed at the boom box; and firing at it twice, blew it away. While the entire pistol-firing-blissful-symphony-in-motion was in action, the defiant Marines were rapidly scattering like rats deserting a sinking ship. As the boom box lay smoldering, the Gunny placed his weapon on safe, holstered it, and peering from under the rim of his MP helmet at the disbursed Marines said, 'End of subject.' Then he turned and walked toward his Jeep while the defiant Marines just stood there in amazed disbelief.

The night may settle some arguments, but the dawn always brings a new twist to old endings, and so it was with this pistol-whipping event.

After 0800 muster the next morning, Mick's Platoon Lieutenant directed him to stand fast after fall out. When doing so, and after the other Marines had moved on, the lieutenant instructed Mick to report to the Old Man's Office immediately. The term of 'Old Man' is slang for the Unit Commanding Officer (CO). Once in the COs office, the CO's aid directed him to stand outside his door until summoned. Within minutes he was; he entered, rendered the proper military courtesies, and stood at attention. The CO tossed two spent .45-caliber pistol casings onto the top of his desk. Looking up at him, the CO said, 'Gunnery Sergeant Mick, you want to explain how these two spent .45 shell casings came from your pistol and into my possession, please?'

'Well, sir, I was on MP Duty and answered...'began the sergeant's discourse before being interrupted.

'Yes, Sergeant, I know all about that. Skip over it and get to the most important part just before those two bullets left your gun' said the CO.

'Sir, the men mocked my rationale to Taps and rest. And, the boom box owner blatantly defied my advice for him leavin' the unit turned off or its volume lowered. He switched it back on to its previous volume. Therefore, I met their mockin' challenges, and his rebellious challenge, with one of initiative that would permanently settle the issue for the evenin' sir' responded Mick.

The CO rose to his feet from behind his desk and said, 'I was not informed the complainant switched the box back on after being advised to leave it off or lower its volume. Neither was I informed of their taunting of an MP. Sergeant Mick, can't say I'd done any different myself if put in your boots with this situation. However, because there was a firing of a weapon, for sake of Official Inquiries, I'll have to dispense punitive judgment against you in the Complaint. I fine you for the cost of the two rounds of ammo. You are dismissed, Gunny.'

'Yes sir' was emitted from the sergeant's surprised mouth. With smiles between both Marines, proper military courtesies were exchanged, the sergeant did an about face, and departed the COs office. While Gunny Mick was assigned at that Marine base, loud boom boxes were never a problem ever again – before or after Taps.

There is a valuable life lesson in this true-life story. Get good and proper rest when possible. Take proper care of this body. After all, we are only issued one per life-time. When it is gone, there are no spares in the supply system. There will always be those who sneer at those who try taking quality care of this body, and this life as well. For those who defiantly sneer at others in the caretaking modes of this life, there are sentinels on duty who will shoot holes in their destructive efforts.

Those sentinels of this life are the professionals that God blessed with high intelligence for protecting the rest of His people. They are the doctors, lawyers, engineers, clergymen, police officers, et cetera.

On the spiritual flip side, God has His angels who are Heavenly Sentinels keeping guard over His people day and night. God's Holy Spirit also flows everywhere to be with those who call on Him in emergency situations. His Spirit and His angels are the Guardians keeping watch over all that we do. God and His angels have no problem whatsoever defending His people against those who are bent on destroying them. His defense actions, though, are in His time, not ours. Because God loves everyone, He takes gracious time for reasoning with the defiant ones. God is always slow to anger; sometimes His actions can be quick; and His judgments are always sure. However, when God sees rationale is becoming totally wasted, and His people are becoming devastated, He can become somewhat like the Gunny. When evil's tide comes rising high against His people, He *can choose* for taking swift action against blatant defiance. He *rarely does so*, but when He does, corrective measures may be very uncomfortable and may come very quickly.

'The LORD executes righteousness and justice for all who are oppressed" (Psalm 103:6).

"The LORD is merciful and gracious, slow to anger, and abounding in mercy" (Psalm 103:8).

"Then Jesus went into the temple of God and drove out all those who bought and sold in the temple, and overturned the tables of the money changers and the seats of those who sold doves" (Matthew 21:12).

"For the Word of God is living and powerful, and sharper than any two-edged sword, piercing even to the division of soul and spirit, and of joints and marrow, and is a discerner of the thoughts and intents of the heart" (Hebrews 4:12).

Classified?

Buried deep within the functional layers of some military and government agencies are stealthy-to-covert operations. So safeguarded are some tasks, even some of their own agents perform select assignments on a need-to-know basis. Those responsibilities require security clearances of 'Classified' and higher. Those clearances permit agents access to select sensitive information. Yet, over the course of events, certain peculiar clues can sift through the hour glass of time, and then, rising to the surface, strange images can appear.

Such an unusually surfacing image 'might' be as the unusually odd one above.

Once upon a time, I was assigned to a unit at a military base (which will not be disclosed). The unit's command center was located in an old renovated building from the WWII era. Not long after reporting in, my wanderings met me with a profound surprise. One of the hallways at one end of the building contained colorful murals of creatures like in the image at the beginning of this entry. At one end of the hallway was a metal placard stating the wall murals had been painted by POWs during WWII, and that the creatures were from the POW's ancient folklore. The placard further stated the base commander of the early 1950s decreed this wall should never be defaced or destroyed. While assigned to that base, I discovered the base would periodically maintain the wall murals for retaining them as close to their original status as possible.

Some months passed, and I was able to snap a few photos. Years of researching that particular ancient folklore followed, and failed; for no comparable matches to those images on the wall ever surfaced. My thoughts clashed over this, for why maintain a wall mural of several unusual creatures from ancient foreign folklore – especially since they apparently did not exist in myths. Adding to this mixture were some events which began creating a mural madness about this mysterious issue.

Just a few years prior, I had been assigned to a unit where a soldier related a most interesting story. In the early 1950s, Jason was working on his private pilot's license and had gone on an unforgettable cross-country flight. His flight instructor was from Dayton, Ohio, and wanted to fly there to visit family for a few days. Since Jason had always wanted to visit the Wright-Patterson Air Force Base (WPAFB) area, this cross-country flight would satisfy several wishes and requirements in one trip.

The flight to Dayton was a good one; and after landing, the flight instructor's relative was there to pick them up. The next afternoon, Jason and the flight instructor borrowed the relative's truck to go sight-seeing around the WPAFB areas. By evening, things became not so good around some base areas; traffic had become very congested. Finally, the police began routing drivers out of one particular base area. Jason and the flight instructor had to take a northerly route around Dayton in order for navigating back to the relative's home. They discovered the next day before their return flight that a train had derailed in the Dayton area. The derailment was what had generated the traffic jam the evening before, and the rerouting of traffic around certain areas of Dayton.

In 2011, a movie was released which was written and directed by J.J. Abrams, and produced by Steven Spielberg – it was called *Super 8.* Even though it was a science fiction movie, my mouth fell open over parts of the theme and its scenes. True, there is no Lillian, Ohio, as was portrayed in the movie. However, some of the events that Jason described when going through their traffic jam paralleled similar events in the movie *Super 8*. Therefore, connecting some *'significant'* dots within *Super 8*, Jason's cross-country flight, and the wall mural – those dots might reveal something:

- Timeframe for the movie *Super 8* was the early 1950s.
- Jason's cross-country was in the early 1950's.
- The wall mural dedication placard was dated in the early 1950s.
- The base commander had decreed for the murals never to be defaced. Why would a high-ranking official decree those murals of ancient folklore always remain intact, especially when comparable images within the folklore cannot be validated?
- The images on the wall murals have an uncanny resemblance to an extraterrestrial alien portrayed in the movie *Super 8*.
- The building where the wall murals are located has become increasingly more secure and thus, is less open to the general cadre of troops.

Applying some additional logic, sometimes leaks happen with the handling of classified material and events. When this happens, special agents are dispatched for corrective measures. Those measures can range anywhere from dust-off to erasure. In between, though, there can be quite a clean-up job with corrective actions ranging from collateral damage assessment to complete stoppage of events.

As with any covert operation, *deception* can be just one of many effective tools for minimizing damages from leakage, as well as incoming barrages. Fantastic wall murals and great movies just might be very effective tools for deception. But then, how would I know? Most of my training came from the military. I've learned to collect facts, try being logical, and attempt connecting a few scattered dots. In doing so, it's surprising what those dots can bring into view – sometimes a much larger picture not usually seen.

This story was built with the facts as I found them, linked with threads of objective conjecture, and laced with the logic a reasonable person would apply. At the time of publishing this book, only its readers can conclude this story's actual connectivity to reality. Regardless of whether one believes in aliens or not, there still remain the indelible facts of unusual and unexplained events occurring on this globe. In many instances, there does appear to be cover-ups instigated by higher agency officials and the media.

Some believe alien extraterrestrials exist, some do not, and others wander in that no-man's land in between (that of not really deciding). If alien extraterrestrials DO exist, a large number of world agencies are engaged in great extremes for covering up the surmounting evidence. If alien extraterrestrials DO NOT exist, those same agencies are STILL sure going to great extremes for covering up the hoaxes. And WHY would be a serious question for either of those possibilities. The truthful answers for those two questions may take many long decades for uncovering.

In view of this unusually odd story and establishing the need for collecting sound facts, the stage has been set for leaving this 'military twilight zone' and entering another fragmented one. The next zone of defined uncertainty lies within the bounds of the next true story.

It has long been declassified, so enjoy, and read it well...

Military UFOs?

Once upon a time, at a small military airfield, an unusually strange aircraft made an emergency landing. Upon landing, it was maintained at the far end of the runway. Several ground support vehicles immediately began surrounding the unidentified strange craft. The surrounding cover-up was so completely thorough, only the top areas of the aircraft were visible. The aircraft was only on the runway for about ninety minutes and then was airborne again (so it was rumored).

Immediately as the aircraft was landing, the base siren sounded off the appropriate warning for mandatory immediate section muster by all personnel. Base Military Police (MP) began appearing in combat gear with armed weapons for expediting the process of mandatory personnel muster. Having practiced this drill many times, within only a few short minutes, all base personnel were in their assigned point of muster for their occupational specialty. With this mustering exercise, however, there were two differences.

Our section meeting room was guarded, as usual, by two MPs; but this time their weapons were loaded. The section chief wore his steel pot (combat helmet) and sidearm as usual, but this time his sidearm was loaded. Beside him was an MP, and his weapon was also loaded. The section chief took roll call, and then the outside MPs slammed the door shut, bolted, and locked it from the outside.

With a solemn face, the section chief stated, 'Men, this is no drill, this is the real deal. Lips sealed and listen until I finish. Some of you think you saw some strange aircraft landin' on the airfield here today. You did not. Some of you think you saw a laser gunship. You did not. What was seen today is an aberration from a Hollywood movie being made. IS THAT PERFECTLY CLEAR?'

'But, Chief Sarg—'

With his hand resting on his sidearm, the section chief expelled through his teeth, 'Did ya all not JUST HEAR WHAT I SAID? Lips sealed until I'm done talkin' and I ain't done talkin' yet. The next man that opens his trap before I'm done will be court-martialed. NOW IS THAT CLEAR?'

A pin could then have been heard dropping in the room as the section chief continued, 'If anyone on this base ever says anything about an aircraft landin' here today, it a laser gunship, and it NOT a Hollywood movie prop – that individual WILL receive a court-martial. The only way out that individual will have is declassification. THIS IS CLEAR TO EVERYONE.' I am finished, I am not taking questions, and the MPs have a form everyone MUST sign before leaving this room. This IS CLEAR TO EVERYONE IN THIS ROOM.'

And so, in total silence, every man in that room read and signed his form; there were no duplicate copies, only originals. Even after all forms were signed, we were detained in the section room until released by the armed MPs. Resuming our duties, the first thing most of us did when going outside was look toward the end of the runway. Nothing was there; all evidence had been 'dusted off.'

Once upon a time, when assigned to a large military airfield, I was leafing through a military magazine. The magazine was the official publication for one of the four military services of the USA. Turning a page, the headlines glared at me, 'United States Air Force Unveils Laser Gunship,' and below the caption was an aircraft photo. The aircraft in that photo appeared exactly the same as the unusually strange aircraft which had made an emergency landing many years prior at the small airfield I had been deployed to.

Before putting the magazine down, the entire article was digested. That laser gunship from many years ago was an experimental highly modified 747 airframe. It contained an on-board Airborne Laser (AL) and was one of several in the U.S. Air Force inventory. The magazine in my hands became public declassification from secrecy of that weapon system.

Today, much more advancements to laser gunships have been made. The Advanced Tactical Laser (ATL) installed in a modified AC-130H aircraft can fire its high-energy laser through the Beam Control System (BCS). The BCS acquires ground targets and guides a laser beam at those targets, as directed by the ATL Battle Management System. By firing the laser through the BCS, the ATL has proven high levels of sustained dependability.

The accuracy of today's laser gunships is little short of supernatural. Their pinpoint precision strike capability can make them lethal or nonlethal at will. For example, they can either destroy a vehicle completely, or only damage the tires for immobilizing it. At an unspecified attitude, in one 26-second engagement, one laser beam can destroy 32 tires, 11 antennas, 3 missile launchers, 11 electronic warfare devices, 4 mortars, and 5 machine guns – while at the same time avoiding harming a truckload of refugees and all the soldiers guarding them. That is selective precision covert destructive capability in the highest degree.

Why have I written about these laser gunship escapades?

They were around for who knows how many years before becoming declassified for the public's knowledge. The 'star wars' technology used within those laser gunships, may be what some people have seen as strange lights in the night skies. Who knows what else exists out there in advanced technology land that John Q. Public will not ever be made aware of for many years to come. *Thus, many of the UFO sightings are most undoubtedly planet Earth originated.* But some of the UFO sightings just cannot be explained away so easily.

There may be another explanation, and it just might be in the spiritual realm – a dimension no one fully understands.

Think for just a moment about God creating things. There is an infinitely huge universe out there. God is an infinite being and He loves diversity. Reading the Bible reveals those attributes. Why would God limit creating mankind to just one place in all the universe – a sphere called Earth? It has always been my opinion He has other people scattered out there somewhere, and they are possibly more advanced than we. Maybe they visit this sector of the universe occasionally. Because satan fell from Heaven eons ago (reference Luke 10:18 and Revelation 12:9) and landed here on Earth, maybe residents from other places in the galaxy have a 'non-intervention directive' similar to that in the *Star Trek* television series.

That concept would bring up the topic of alien abductions. Personally, I think they are all hoaxes. If there were anything to them, though, demonic activity would be a serious consideration in my opinion. Why? In Revelation Chapter 13 verse 13, it is mentioned, *"He performs great signs, so that he even make fire come down from heaven on the earth in the sight of men."*

The 'he' in this passage of Bible Scripture is what Bible scholars label the False Prophet. He is the right-hand man for the Antichrist, the devil incarnate during the most tumultuous period of time this world will ever know. From reading Bible prophecy and watching the media, that time may not be too many more years into the future. So then, if the devil and his works can perform signs in the sight of people, even fire in the sky, then that leadership from Hell can do smaller works now. Those hellions could wreak havoc with what people think about their origins, and who/what is surrounding them.

As mentioned before, UFOs were at first primarily considered being not of this world. And again, why do we believe such unknown, unexplained events are not of this world? We *choose* in believing it because we do *not* want to be *alone* in life. We do not want to be alone in the solar system, the galaxy, or the universe.

Not one of us wants to be alone.

Considering this further, why couldn't it be possible that God does NOT desire being alone in His universe and that is why He made us, His prize creation? Further considering, because we have rebelled against Him and have gone our own ways, He may have established angels for watching over us all the days of our lives. Those angels would be 'Guardian Angels.' Just maybe some of the lights, strange phenomena, and other oddities seen in the skies are angels and demons fighting over God's prize creation (us).

Yes, that concept may be new (I've never heard it before), and will sound very strange to many. I am most sure many of my Christian friends (and secular ones also) will ridicule me for this opinion. However, in support of this opinion, I have as a possible explanation for 'some' UFOs, seriously consider the following four Bible Scriptures:

- *"...from the first day you set your heart to understand, and to humble yourself before your God, your words were heard; and I have come because of your words. But the prince of the kingdom of Persia withstood me twenty-one days; and behold, Michael, one of the chief princes, came to help me, for I have been left alone there with the kings of Persia"* (Daniel 10:12-13).
- *"For we do not wrestle against flesh and blood, but against principalities, against the powers, against the rulers of the darkness of this age, against spiritual hosts of wickedness in the heavenly places"* (Ephesians 6:12).
- *"Are they not all ministering spirits [angels] sent forth to minister for those who will inherit salvation?"* (Hebrews 1:14).

- *"Do not forget to entertain strangers, for in doing some have unwittingly entertained angels"* (Hebrews 13:2).

The Bible mentions the angels of God many times. On some of the occasions, those angels conduct guardian or ministering types of duties. The Bible also mentions demonic activities many times and their intent on destroying mankind. With the previous mentioned four Bible Scriptures, it is very evident of a spiritual warfare existing, on this sphere called Earth, between good and evil over mankind. It could be that some UFO movement is the trailing edge of warfare activity in the spirit realm for salvaging our souls and saving our Earth.

Such is my opinion anyway…

Dangling IO

In my flying days as an Aerial Gunner, all flight crew members periodically received aviation safety briefings. These briefings were conducted by the unit Safety Officer and, at various times, were augmented by other unit members. The sessions always contained materials directly related to our flying weapons platform (the aircraft we flew in). The safety briefings were for bringing awareness to mishaps occurring in the field and for us avoiding the same, or similar, mistakes. Some sessions also contained videos for enhancing the applied safety topic.

In one safety briefing, we were informed of a malfunctioning tether reel. In another squadron like ours, a gunship had been on a tactical sortie. While in live-fire, the Illuminator Operator (IO) was in his flight position for combat, that of literally hanging half way off the gunship's back cargo ramp. His job was observing for incoming anti-aircraft fire and missiles. In this position, the IO is secured to the aircraft by a metal cable – a steel tether line. The tether line is connected to a tether reel, which is similar to winches on 4x4 trucks. The IO has a remote which controls the winding and unwinding of the tether reel, and its locking and unlocking operations.

While in the live-fire operations with the aft cannons of the AC-130 Gunship, the ratchet mechanism of the tether reel inherently malfunctioned into the Release Mode. The IO was suddenly released from the aircraft cargo ramp and into the air! Several feet of steel cable later, for some reason the tether ratchet mechanism reengaged and locked. The IO happened to then be deployed at about the area where the right and left wind slip streams converge underneath the aircraft tail section. This resulting wind turbulence was wreaking havoc by slamming him against the underneath and sides of the aircraft tail section.

When the aft aircraft cabin Aerial Gunners saw this, they immediately called 'Cease Fire!' and reported to the Flight Engineer an In-flight Emergency. All guns were placed in the Safety Mode while the Flight Engineer was informed of what had happened to the dangling IO. The pilot maneuvered the aircraft out of orbit and leveled off (in short, orbit is the term for flying at a specific altitude and angle around the target). The gunners then began manually pulling the IO back into the aft cabin of the aircraft, until they discovered how to reset the tether reel, re-engage it, and then reel in the injured IO.

Once inside the aircraft, the IO was found bleeding, semi-conscious, and in need of first aid; which he was given until the aircraft could make an emergency landing at the nearest airbase. The IO survived the slamming event, but sustained multiple broken bones and lacerations. Several months later, his unit's Flight Surgeon certified him fit for resuming his aircrew duties. Because that IO loved *the red, white, and blue* (and flying so much), he continued to fly again in his same aviation flight position as an IO.

There is actually a spiritual moral to this dangling live-wire story – *God watches over His people when they are not aware.* His Holy Spirit is like a 'tether' when we have been released into the free falling slip stream of the chaotic world. That release may be accidental, or intentional. And, we may even have *intentionally* cut the security cord between us and God. Yet, God loves us so much the 'reel' of His real love defaults and *kicks* in after we have played out into the danger zones of this life. His 'real love' locks securely on to us and WILL NOT let us go. Then He sends His Guardian Angels around for pulling us back into safety again.

Tethers of this world ***WILL*** fail, but God's anchors always hold – on land, air, and sea! (reference Matthew 11:28, John 3:16. John 10:28, and there are many more).

The view AC-130 Aerial Gunners (AGs) and Illuminator Operators (IOs) have from the aft cargo ramp just prior to taking off from the runway.

The Voice of God?

'But you really don't remember, was it something that they said? Are the voices in your head calling, Gloria?' Recall those lyrics from the song *"Gloria"*? The song went to number two on the U.S. Billboard Top 40 Charts in 1982. Italian pop singer Umberto Tozzi originally recorded the song as a ballad in 1974, but Laura Branigan sang its most popular, best-known version.

Thinking about 'voices in our heads,' how do we know when the Lord God impresses us with His Words or His thoughts? How do we know that He has actually 'spoken' to us or 'spoken' into our spirit from His Spirit? The answer to that towering question is actually a personal one. The Lord God connects with each individual in a personal way. The way He communicates with me may not be exactly the same as the way He communicates with others. Yes, God does speak audibly, but it is rare. In the Holy Bible were Prophets in antiquity who were devoted to following God and communing with Him. Because there was little to no written Word of God in those days, God chose the use of Prophets as His mouthpiece. The Bible also records where God used angels as His messengers for carrying His Words.

How then does an individual know the voice of God impressed into their heart, mind, or spirit? The follower of God through Christ must follow in a close relationship. Best friends sometimes know each other so well they know what each other will say next because they know how they think. This is because they are in each other's presence so much of the time, and that familiarity forms some bonding of shared mind and spirit characteristics. That principle works the same with the Lord God and us. Then when God speaks, or impresses something of

Himself into us, we sense it, and know within us that it was from Him.

God spoke to the Prophet Elijah once when Elijah was going through a difficult time with being threatened by Jezebel (reference Chapter 19 of I Kings). God told Elijah that after all the winds came and tore into the mountains, the rocks had broken, an earthquake came, and a fire erupted, He was not in those things. After all those tumultuous things, He was in the still, small voice, which followed them. God's voice and directions are usually not the loudest and most salient things we hear and/or observe. He is this way because He wants our spirit to be in contact on a regular basis with Him so that regular familiarity identifies Himself to us through all the clutter when He speaks.

Consider again 'all those voices in your head' from the lyrics of *"Gloria."* What if there were ways to project voices into the heads of people? What if the technology existed, or was soon to exist, which could project thoughts into the minds of individuals? How would individuals know if thoughts were actually their own originated thoughts or implanted thoughts? Considering this further, how would individuals know the voice of God or impressions from God?

Consider a military weapon called 'The Voice of God' or 'The Voice of Allah.' Conducting Internet searches on this technology will reveal very surprising results. Those results are that the weapon does exist, does not exist, is in developmental stages, and/or is very speculative. The results can be found scattered in editorials within *The Washington Post* to *The Star*. The Voice of God Weapon is a device apparently being developed by the military and operates at a distance, delivering a message in which only a single person can hear. The messages are in effect that of 'Stop the bloodshed,' 'Surrender,' or some other command to the benefit of the hearer's opponent.

During the development of this book, there have been recent technical workshops reporting The Voice of God Weapon has successfully been field-tested. Results indicated that one insurgent in each subject group would begin looking in all directions and then engage controversial conversations with his companions – those who did not hear the 'implanted' message. The purpose of this weapon is the disruption of thought processes of the opponent. The end result is confusion within the opposing troops and beginning the breakdown of communication within the chain of command, thus, rendering opponents useless and defenseless.

As with all weapons used for defensive purposes, they can also be used for offensive strategies as well. This same weapon in the wrong hands could have devastating potential, and one of those is with terrorism. Adequately funded fanatical terrorist factions could use this weapon to their advantage against good peoples and nations with well-meaning societies. Just ask any good soldier, sailor, or airman from field, and it will be verified that any weapon has good and evil potential. The end result of any weapon is the intended goal.

Having considered intended weapon goals, look further into the Voice of God weapon falling into the hands of a one-world leader. With the advent of global transportation and communication in ever-reducing time frames, the world is becoming 'smaller' socially. This current world is shaping up politically, economically, socially, and religiously, for one government to solve all of its global problems. Several hundreds of books have been written about the possibilities of one-world governments and leaders. With nightmarish problems impacting the global community, these problems become more and more of world concerns by world citizens. Therefore, one leader could arise and make a difference among all of them – in theory.

What if a leader in the world's celebrity political scene did arise in popularity? What if that one leader possessed the charisma to sweep the world beneath their feet? What if that one leader was given leadership authority over all governments of the world? Then, what if that one-world leader became corrupt and perfected the Voice of God Weapon for a broader-scale application? The regime of this leader could use it to program the citizens of the entire planet Earth. Programming could exist at voting stations, licensing bureaus, application booths, advertisement stands, shopping centers, etc. The list could literally go on and on. The most obvious places for programming would be the places of employment, especially federal and state governments. Secondary places would be through the media of radio and television (when the technology is enhanced).

Now, enter the Antichrist. The Holy Bible also calls him the Beast in several passages of Scripture. This man is the epitome of evil because he is actually satan incarnate. He will imitate everything that Christ did in order to confuse everyone who believes on the one true living Lord God Almighty. For many years, it has been believed the Antichrist will have the ability for instituting a global numbering system for tracking everyone. Numbers issued by the Antichrist regime will be indelibly imprinted on the right hands or foreheads of every person and will track the peoples of earth (reference Revelation 13:17-18 and 14:11).

The Antichrist will also have the authority for enforcing all citizens of planet Earth to worship the Image of the Beast. This is an image of himself, and all who do not worship this image will be put to death. This is found in Revelation 13:15, *"He [Antichrist] was granted power* [by satan] *to give breath to the image of the beast* [Antichrist], *that the image of the beast should both speak and cause as many as would not worship the image of the beast to be killed."*

What is to say that The Voice of God Weapon technology could, or could not, be modified for use in brainwashing techniques of mental reprogramming? If the one-world leader in charge wants it done, who is to stop the motion, especially if that entity has god like powers? What if The Voice of God Weapon technology were used to enforce submission to the new one-world regime and especially its leader – by way of worship? And that worship just might be at the television, where a hologram of the Beast could someday be projected.

These ideas may be a little far-fetched, but what if I am not wrong? What then? How will you know the Real Voice of God versus The Voice of God Weapon technology implanting something in your head? These concepts ought to be seriously considered, especially since The Voice of God Weapon technology is a reality, despotism in the world is a reality, and there is a real devil out there in reality. So then, how would you discern the difference between the real voice of the Lord God and the false voice of god technology?

Christ Jesus spoke much in Chapter 10 of John about His people knowing and hearing His voice. Then He said in John 10:27-30, *"My sheep* [His people] *hear My voice, and I know them, and they follow Me. And I give them eternal life, and they shall never perish; neither shall anyone snatch them out of My hand. My Father, who has given them to Me, is greater than all; and no one is able to snatch them out of My Father's hand. I and My Father are one."* Christ was stating that if we know Him personally, we can hear His voice when He projects it within us. Whether His voice is in a dream, a vision, an impression, a sensation, a thought, or an actual voice – we will know it came from Him because we know Him.

Then later while Jesus was on Trial and talking to Pilate, He said, *"...Everyone who is of the truth hears My voice"* (John 18:37). One must seriously consider something here with this passage of Scripture. When coming to God through Christ Jesus, repentance of sins is required. This is being sorry for the wrongs committed and wanting to turn away from them. Then with believing Christ's sacrifice on the cross as payment for sins and accepting it, a regeneration of the inner spirit begins with the rebirth of the soul. That regeneration process is a continual process until death comes.

That regeneration process can be fast, slow, shallow, and/or deep. It depends on the interest of the individual. The deeper the interest will usually generate a faster process. Additionally, the deeper process will bring the individual into a very close personal relationship with the Lord God. The deeper the relationship, the more truth and honesty is desired and earnestly sought after. In fact, the deeper the relationship, the more truth and honesty are hungered for. The more hunger for truth and the more in communication with God (talking with Him), the clearer becomes the sound of His voice. Thus, the dedicated Christian will have no difficulty whatsoever distinguishing the Lord God's voice from any confusing voice of god weapon technology.

Discerning the Lord God's voice, at such a time of this twenty-first century with all its technologies, is truly a personal impression from God; and God knows how to imprint Himself into each and every one of us. We need to earnestly seek the encoding processes. We also need living a life close to God, stay connected to His Spirit, and remember Isaiah 54:17, *"No weapon formed against you shall proper...This is the heritage of the servants of the LORD..."*

God's Tech Manual

All during my times in the military, everything I did was accomplished through some manual, with supporting directives. In the maintenance community, those manuals and directives were published and issued by what was generally called the Cognizant Field Activities (CFAs), Depot System Commands (DESCOMs), Air Logistics Commands (ALCs), etc. CFAs obtained their technical data directly from the Original Equipment Manufacturer (OEM). The OEMs were staffed with engineers of various skills who designed and built the military equipment, and then tested their limits to sustain the operational demands needed by the various field activities.

When performing maintenance on military aircraft, I was to perform those maintenance tasks exclusively by the Aircraft Maintenance Manuals specified by the CFA or for that particular Type, Model, and Series (TMS) of aircraft. Any deviation from those specific 'tech manuals' would most likely result in an improperly maintained aircraft. That, then, would result in a very poor quality weapon system, thereby degrading fleet readiness and thereby degrading national security. Improper maintenance would also most likely result in an aircraft mishap, resulting in millions of dollars of weapon system loss, and unknown numbers to loss of life. Therefore, there are many reasons for following tried-and-tested maintenance and repair procedures by the book. No other tech manual would be applicable for the unique systems in the aircraft TMS, which I was maintaining or flying in.

Such is similar with the Bible. Someone once said and wrote that the letters B-I-B-L-E could stand for 'basic instruction before leaving earth.' That does sound excellent, especially to those who come from a 'tech manual' background. God inspired several men, through His Holy Spirit, to write the Bible, the Book. God is the Creator of the universe. He made it all; He knows how it all works. He made mankind, and therefore, God, as the OEM of everything, knows more about the original design of mankind, inside and out, than any of the rest of us. God's detailed, intimate, and thorough knowledge of humanity is past, present, and future.

Doesn't it, then, make logical sense that the Bible be consulted for guidance regarding the issues of life? Why not go directly to the writing of humanity's OEM for the answers to all the concerns of life? The Bible contains lessons on national as well as individual levels, yet applicable for any level of society and culture. The Bible contains answers to easy issues and hard ones as well. The Bible is full of history about peoples making some serious mistakes and rejecting God's help. The Bible is full of history about peoples making some serious mistakes and accepting God's help. The Bible shows the results when we 'do it our way,' and it shows the results when we do it with His help, His way. The Bible is a tech manual, an operator's manual, and a guidebook, which God gave to the people He made – us. The Bible is also God's love letter to His prized creation – us.

Precious metals and stones have to be mined, and that mining comes with patience and diligence. So do answers from the Word of God; they must be mined with diligence and patience, for it is more than just any book. The Holy Bible is a spiritual guidebook: *"Your word is a lamp to my feet and a light to my path"* (Psalm 119:105). As the answers come and are revealed to the patiently diligent seeker, those answers sometimes become so profound their revelation becomes imprinted within the heart and mind. Those imprints are personal imprints for that one who sought truth from the Lord God. No one can then steal those imprints away from the heart of seeking pilgrims in their journey of life.

With those precious imprints, that one will always have a map for traveling the road of life, a technical manual for navigating its difficulties, and even a construction manual for expediting the repairs with the walls we drive through scattering shrapnel everywhere!

Best Letter Ever

Once I had a friend by the name of Rob who worked in an office complex next to mine. He was a retired U.S. Army Intelligence Officer and had gone to work for the U.S. Air Force after retirement from the Army. Rob and I discovered some slight similarities in our military careers, which then lead to us taking breaks together. We'd swap war stories during those breaks on the loading dock at the back of our office building. The day came when Rob discovered of having a brain tumor, and it was of the fast-growing species.

The time came when Rob had surgery for removing the tumor and, within a few weeks, returned to work. As the attitude of many career military personnel would be, his was resilient and still sharp and eager for reengaging himself again. His reengagement was short, however. Only about six months later, Rob discovered his tumor was returning.

This time Rob was not given a good prognosis. Rob knew he would most likely die within a very short time. The doctors told him it could be a matter of weeks to a few months. It actually turned out to be four months from the time of discovery of the tumor returning. At this point, having only known Rob for just under a year, I wanted to find the right time for talking with him about God and eternity.

Rob was sort of a 'crusty' older Army Officer whose eyes were filled with battlefield stories from all over the globe. When he told of them, his voice leaked of those agonizing but devoted sentential duties. It was his periods of silence as his eyes wondered off into space, grasping for the present, while reliving a past that told another story – that of a guardian in the outlands.

It was his silence with devotion to country I greatly respected. I did not desire breaking or disrespecting any trust Rob and I had built in our friendship. So I prayed that the Lord God would lead me in finding the right time and avenue for talking with Rob about God and his eternal destiny. The matter was left in God's hands, and only a few days later, the answer came while looking at a magazine.

A military memorabilia magazine had come in the mail, and its front and back covers were a fantastic collage of military items. The magazine was a special issue, which was printed in sharp military colors and high gloss finish. An idea instantly came into mind to write Rob a personal letter, fold and place it in an army green Gideon New Testament Bible, wrap them in the magazine cover paper, and give them to Rob. The letter was written as follows:

Dear Rob,

I sincerely hope this letter does not offend you. From our past conversations, I understand that your medical situation is not good at all. Having just become acquainted with you back in the summer, I just wanted to say that I have grown to think a lot of you. I wish that we could have been friends a long time ago.

I wanted to give you this New Testament Bible. Don't know if you have one. Rob, I have been a Christian for many years. The Lord Christ Jesus has spared me from certain death and harm several times. The Lord has been a very real person to me ever since I asked Him to forgive me of my wrongs and to come into my heart long ago. I have not been perfect and have made a lot of mistakes during all those years. But the Lord never turned away from me. When someone cares that much, you tend to stay with them in a deepening relationship as time goes on.

In your past military and government career, I believe there may have been many times that God has watched over you and protected you in the circumstances you have faced. I ask as a concerned friend and Christian—would you consider inviting the Lord God to come into your heart?

I had an uncle that I thought very much of. Uncle Stan passed away a few years back. He was a highly decorated WWII veteran. About two weeks before Uncle Stan passed away, Aunt Jane woke up one night and saw an angel at the foot of their bed. The angel then turned to exit through the bedroom window. Aunt Jane spoke of this to Uncle Stan, and a few days later, he woke up one night and, while lying in bed, asked God to forgive his wrongs and invited Christ Jesus to come into his heart and life.

My Aunt Jane said that during the next several days of Uncle Stan's life, he was the happiest she had ever seen him. She said he even walked up and down the neighborhood telling their neighbors about the Jesus he came to know, and they should come to know Him also.

So in closing, Rob, please read the inside of the back covers of this New Testament Bible. It shows that God sent His Son Christ Jesus to die for us so that we can be restored into fellowship with God. All it takes is asking and believing. I ask as your friend that you consider this. Christ Jesus is the best thing that ever happened to me in all my life. If you ever want to talk about the Lord, you can call me any time. I would be glad to talk with you.

Your friend,
Carl Adams

It was a couple of weeks before the specially wrapped present could be given to Rob because of him being ill and out of the office. When he did return, he was not feeling well, but he was very surprised when I gave him the small package. My instructions to Rob were not opening it until he got home from work that day.

The next time I saw Rob was a couple of weeks later, and he was not well. He did not speak of his small package, but he really did seem to be enjoying talking with me more than before while on breaks. Even though Rob did not feel well most of the time, he would perk up when he saw me walking toward him. Rob was only at work another month, but only a few scattered days during that month, and partial days at that. Actually, I only saw Rob four more times before he retired the second time.

Concern swept over me that I'd offended him in some way. However, I did what I'd felt impressed for accomplishing in witnessing to Rob about God and eternity. There are times, though, we can 'feel' that our actions were right, but yet not have any sensation of any comfortable immediate outcome. Those are the times faith must be the agent for carrying one over the rocky roads of reality.

Then came a day in the cafeteria, a mutual friend informed me Rob had died. She lived in their neighborhood and had attended his funeral. She told me that Rob had requested the song *"Amazing Grace"* be sung to him at home during the week prior to his passing away. By that action, I assumed he did have an interest in God, but did not know how much.

Months passed, my curiosity would not rest, and I asked the lady for Rob's wife's phone number. Upon calling Cheryl (Rob's wife), my identity was announced and I related the reason for my call. I stated of having been wondering so much if Rob had ever mentioned anything about the Bible I had given him. The response was a tremendous surprise, one I'll never forget!

Cheryl told me she would never forget the day Rob came home with that little package I'd given him and the following events surrounding it. She related Rob would usually come home and sit in his easy chair and catnap until she fixed dinner (because his fatigue was so great). That particular day, Rob came in, sat down, and immediately began opening the package. She came over and asked him about it, and he related a friend at work had given him a small present.

Cheryl then went on to finish preparing dinner and later heard Rob crying. As she came to him, he was bent slightly over with the letter and Bible in his hand. She asked him what was wrong. Rob related nothing at all was wrong, only that he had just been given the best letter ever in his entire life.

Cheryl went to say that after dinner, Rob sat in his easy chair and read that little Bible for hours. He would also reread that letter, and tears would come to his eyes. He would do this over and over again until he passed away. She also told me that because of that letter and his reading that Bible, Rob asked God to forgive him and for Christ Jesus to come into his heart and life. Cheryl said his last days were his best days because of (her own quote) 'his knowing God, and having Christ Jesus as his personal Savior.'

I sincerely thanked Cheryl for her time, she thanked me for mine and my interest in Rob, and we terminated the phone call. Then I broke down and cried. My tears were for finding out my good army friend Rob made his way to Heaven, and I was privileged to be a very small part of him finding its pathway. My tears were because one day in God's Heaven, Rob and I are going to sit down for a longer period of time and swap more stories. It is my strong suspicion some of our stories will be about amazing grace for being found from a state of being lost.

Rob was in his late fifties when he came to know God before his passing from this life. God most likely has watched over Rob in the many battlefields of numerous outlands, and Rob was never aware of His Divine Protection. God wanted Rob, even if it was at the close of his life. God wanted Rob, and He used me, a piece of wrapping paper, a letter, and a Bible for arresting Rob's attention. No matter where we go, what we do, God wants us. And He wants us so much His love will come after us again and again and again.

Considering the letter I wrote to Rob, there are other letters which need consideration – those to our friends and relatives. If not a letter, then maybe just a note to tell them just how much they mean to us. Those notes and letters can change the course of lives in directions we never thought or even dreamed possible.

Considering the letter I wrote to Rob, there are also other letters which need consideration – those of the Bible. All that is written in the Bible are actually God's love letters to His prized creation – us. He used mankind by Divine inspiration, for scripting the Bible to show us His love, care, wisdom, and guidance, all mingled with history, prophecy, and literature. The Lord God Almighty actually wrote the Best Letter ever – the Holy Bible.

GLORY
DAYS

Glory Days

'*Glory Days*' is a short novelette (pages 171 through 222 within this book) and is about an American GI experiencing WWII in the South Pacific Islands. The works in this novelette are meant to be a true representation of historical facts. However, in verifying the stories related to me over the years, some interesting clashes in 'factual' information were found. Even though much of the information was extrapolated from government resources, some of those 'official' facts still disagreed with other 'official' facts at times.

After literally dozens of hours of investigative work, I found it best in reconstructing this true short wartime story first with the GI who was there, and then blend it with the backdrop of the best 'reliable' facts available in my research. All efforts possible were taken for insuring the accuracy and historical facts of this story. Even after nearly seventy years have transpired since these events occurred, my investigations revealed many websites were still changing and updating their listed information. Many of them were stating their sites were continually under development and their works were always in fluid states of progress.

In my own years of military experience, I am well aware military records do not always reflect the exact details of the service member's time in service or career. Military personnel can be placed in a 'temporarily assigned duty' (TAD or TDY) status into another unit and location. If service members do not maintain their documentation of those transfers, and if and when those records become 'easily' lost forever, there remains no official permanent record of any sort for the missions those service members performed.

Myself, my Form DD214s do not list four award ribbons and one metal. Why? I was discharged before the awards were approved and the paperwork never caught up with me or my records. Additionally, there were missions and assigned tasks I completed and was never given credit for, while others were given credit for accomplishing the same mission and tasks. They received awards, but I did not. Why? In those select instances, my superiors did not submit any recommendations for me. Why? I had 'rattled some cages' in the past, or, had 'rocked the boat' as the sayings go. Recognitions, awards, metals, and official documentations do not get accomplished without the official chain-of-command approval.

Thus, by comparison in my research for this story, smiles came to my face when seeing some fractured and disconnected information. Suspicions were that somewhere official documentation was not thoroughly accomplished. Still, though, it is hoped that all who read this short novelette, will not only enjoy it, but after reading through its pages, will also deeply reflect the sacrifices GIs have made for our lives. It is further hoped its pages will possibly cast thoughtful imprints on the welfare of the reader's soul.

The Terrible Ones

When conducting research on military units, results from interesting to the strange and humorous can always be retrieved. As for the United States Army's 32nd Infantry Division, they were, and still are, quite an impressive group of soldiers. Everywhere they have traveled for mission deployment, they have proven their motto of being 'The Terrible Ones' valiant in battle. Their unit patch of a Red Arrow over the years became the prime ingredient for their slogan of 'The Red Arrow Pierced Every Line' – and it still holds true.

The 32nd Infantry Division was originally the Wisconsin National Guard as appointed by the first

Adjutant General of the Wisconsin Territory in 1839, and was later called the State Militia until 1879. The division was organized from elements of the 1st, 2nd, and 3rd Wisconsin Infantry Regiments (126th, 127th, and 128th Infantry Regiments). The 32nd Infantry Division has always been composed of Wisconsin National Guardsmen, except during WWI. During WWI and WWII soldiers from every state entered its ranks.

The 32nd Infantry Division Unit Insignia Coat of Arms contains a red arrow which represents their mission of piercing the enemy lines. At the bottom of the Coat of Arms is the motto, 'Les Terribles.' During part of WWI the 32nd Division was under orders of the 38th French Corps, commanded at that time by General de Mondesir. During one conflict, the French Commander was at the front observing the battle and saw the 32nd Division in action. The French Commander was so impressed with the dedication, determination, and fighting skill displayed by the 32nd Division that he said, "Oui, Oui, Les soldats terrible, tres bien, tres bien!" (Yes, yes, the soldiers terrible, very good, very good!).

French General Charles Mangin (known as 'The Butcher') heard the report and statement from the commander and referred to the 32nd Division as 'Les Terribles' (The Terrible Ones) when he asked for the division to join his famous 10th French Army troops north of Soissons. The general later made the unit nickname official when he incorporated it in his citation for their tremendous actions at Juvigny. Thus, the unit motto of 'Les Terribles', or 'The Terrible Ones' was derived, and has remained with the 32nd Infantry Division and its subordinate elements.

Flat Footed Draft

Robert Adams of Louisa, Kentucky, was a red-blooded American hillbilly when WWII involved his country. Like many of his friends and relatives, he volunteered for military service. He was not going to let anyone from some other country attack us and get by with it – no sir! The U.S. Army was a little select, though, at the beginning of the war. Robert was turned down because of fallen arches, or flat feet as the disorder was commonly called.

That medical classification did not waver his mindset one bit. Two more times he tried volunteering again. And, as with the prior times, the army physicians turned him down because of flat feet. Arguing each of the three times with the examining boards that he was just as fit as most men his age, his reasons were always denied. The army's arguments always won because a flat foot, and especially two flat feet, would wear down a soldier on long marches.

Watching his friends and family leave for service to their country saddened Robert, not to mention he could not follow in the footsteps of his father who was a WWI soldier. Working at loading train cars at the railroad station in Louisa, he would often see soldiers arriving and departing. Thus, Robert resigned himself to the fact he would never enter the war, and on 27 April 1943, he married the woman he had fallen in love with, Pauline Thompson,

About three weeks later, Robert received his draft notice for the U.S. Army. He was to report to the Local Draft Board on 18 May for induction into the army. 18 May was one day before his twentieth birthday. What a combined birthday and wedding present from the U.S. Army! On 25 May 1943, Robert departed Louisa, Kentucky, by train headed for a U.S Army boot camp in Texas. During boot camp, and because of his medical classification, he was designated for training to become a U.S. Army Field Medic. Flat feet may be a problem for marching, but when the war escalated and needed more men, the army began taking men with minor ailments and placing them where their service best suited the U.S. Army.

After boot camp, Robert was transferred to Camp Robinson in Arkansas, and was assigned to Company 'D' of the 105th Training Medical Battalion. While at Camp Robinson, the base utilized some newly acquired equipment for the GIs to contact their loved ones – the 'Voice-O-Graph' and the GIs loved this technology. With this equipment, the GIs spoke into a microphone and their voices were recorded onto a cardboard record about the size of the 45 rpm singles. The record was then ready for packaging in a brown mailing envelope which contained the trademarked words 'Voice-O-Graph' at the bottom on the front of the envelope. Robert recorded some Voice-O-Graph records and mailed them to his wife, Pauline, while at Camp Robinson.

After graduating from Field Medic School in July 1943, Robert was assigned to advance training for Combat Medics. Once this phase of training was completed in late August, he was given furlough to go home for thirty days before shipping out to the war somewhere in the Pacific. Robert didn't know where in the Pacific he was destined, but his orders indicated he was being assigned to the Asiatic Pacific War effort. When leave was over, he was to report to Fort Ord in California to receive his orders of unit assignment – and most likely his unit would be somewhere in the Philippines.

After arriving in Fort Ord and settling in for a few days, his orders for overseas came in. He was being assigned to the Medical Detachment of 32nd Infantry Division – 'The Terrible Ones.' At that time, the 32nd Infantry Division was being rebuilt in Australia. It had suffered a sixty percent casualty rate in Phase One of General MacArthur's Philippine Champaign. Such news was not good in some respects. For survival, that news generated mixed emotions for it was carried with the fame about a unit known for smashing through front lines against all odds in spite of recently suffering so many losses.

Robert attended a special Chapel Service conducted by one of the Army Chaplains for the soldiers departing the USA and heading to fight the war. In that service, he was given a small New Testament Bible. Before entering the army, he had begun trying his best to make some positive changes in his life; he even thought occasional church attendance would help. He was skeptical, though, for he'd seen many so-called Christians (and preachers also), tipping mugs of moonshine and getting drunk. About that time was when he first laid eyes on a fair farm maiden who attended church, he knew he was hooked (and she would later become his wife). That, he thought, wouldn't be so bad since he had already decided on attending church occasionally. With his life now going to be put literally 'on the line' for an unknown period of time, maybe it was time for reflecting on who God really was. Maybe it was time for getting things right in his life for sure. Skepticism, though, especially of the unknown and the unexplained, will always be the greatest barrier toward any progress. God knows this, and so does His adversary the devil.

On 23 October 1943, Robert departed the USA from Angel Island, California, aboard a U.S. Army transport ship and sailed for Australia. Little did Robert know the wish he had wanted about serving his country and was given, would find him in hell on earth.

But, then, there really are no heavens on earth during times of war – are there?

Bayonets and Rings

While crossing the Pacific Ocean, Robert wished he could have given his wife a wedding ring when they were married. The wedding plans were to save their money and buy a set of rings later because it was taking just about all his earnings to eke out a living. Maybe if he made it through the war, he could bring a set of wedding rings home. Right now, though, he had to focus on staying alive. If he did not, wedding rings were of no value to her or him either one. Thinking of his wife, home, the hills of Kentucky, and if he'd ever see them again – he began reading some in the Bible given him before departing the USA. Robert had a lot of soul searching to do on the voyage across the great pond.

On 16 November 1943, the army transport ship landed at the Port of Brisbane, Australia. Before landfall, Robert and the other soldiers were at the ship's railings taking in the Australian countryside as they sailed in. It was beautiful, but it still was not his beloved Kentucky hills. Long before making landfall, the men of his unit were informed that elements of the 32nd Infantry Division in Australia had recently begun relocating to Milne Bay and Port Moresby in New Guinea, and also Goodenough Island. The 32nd was gearing up for an offensive to remove total control of New Guinea from the Empire of Japan.

After making landfall at Brisbane, members of the 32nd were transported to Camp Cable about thirty miles south of Brisbane. Camp Cable was an Allied base populated mainly with about 20,000 Australian and American soldiers. Robert and others of the 32nd were given only a few short days of jungle orientation training in regard to the terrain they would soon be fighting in. On their first day of training, the Combat Medical Instructor informed the Combat Medics he didn't care what they were told back in the USA, or wherever else they came from, if they wanted to stay alive longer, they needed to pay real close attention to what he said. In one training session the instructor loudly expelled, 'If I ever see them white medical arm bands around yer arms in combat, I'll most likely knock yer lights out ma self. Them Japs use them bands for targets and they know if they can take out a medic they can take out many of us others also. So, if'n ya all wanna stay alive a little longer, use them armbands as a turnakite on the wounded. As for them red crosses on yer helmets; cover'em up with anything ya can find. They make real good rifle targets too.'

A few days later, Robert and the others of 32nd found themselves on another army transport ship sailing for the island of New Caledonia. At New Caledonia, the ship was to drop off supplies, take on more soldiers, and then sail to Port Moresby, New Guinea. Shortly after landing in Port Moresby, they were divided into platoons and sent in the field for further training – bivouacking in convoy across the island to Midino (in Milne Bay Province). From there they were transferred to Goodenough Island for more intense training toward an upcoming assault on Saidor, which was further up the coastline of New Guinea.

It was at Port Moresby that Robert and the others learned what the 32nd Infantry Division ate for Thanksgiving Dinner the year before there in the muck and mire around Port Moresby – C rations and chocolate bars in their foxholes. It was also in Port Moresby that Robert came across two items he would come to hold dear to him – two nickels of New Guinean currency. When first seeing those nickels they sparked an idea. The nickels were around the external diameter of an average size ring. Why not use a rifle bayonet, hollow them out, and shape them into a set of wedding rings? The handmade wedding rings would have esthetic value; not to mention yet another reason for keeping him alive while in combat.

In the days and weeks to come in between combat maneuvers, a set of wedding rings would begin taking shape, and a new purpose for staying alive was born.

Saidor Landing

Intell rumors about an assault on Saidor became more intense. This move was part of the continuing Operation Cartwheel of General Douglas MacArthur for isolating the enemy's positions, cutting off their supply lines, and driving them off the island of New Guinea. Once on Goodenough Island, Robert and the newcomers with him immediately settled into their assigned tent quarters and commenced combat training maneuvers.

Day and night, for the next few weeks the training never ended, not even a slight break for Christmas Eve and Christmas Day. This was a definite indicator the move on Saidor was a serious one. 2 January 1944 became the scheduled D-day for assault (which was termed the Michaelmas Task Force). Logistically, time became of critical essence for organizing all elements of the task force; planning for loading and offloading the ships; arranging for naval gunfire support, air support; securing communications and supplies; planning evacuation for the sick and wounded; and distributing maps and air photographs.

Because many of the attached units for the task force were not a part of the main assault responsibility (that of the 32nd Infantry Division), maintaining secrecy became a problematic issue. Nevertheless, even though rain hampered the entire operation, the main elements for the force were disembarked from Goodenough Island in nine destroyer-transports, several LCIs (landing craft, infantry), and two LSTs (landing ship, tank), during the last days of December 1943.

Training never fully prepares the trainee for the real deal. All the programmed instruction, all the carefully laid out policies, all the best guidelines rendered, and all the repetitive actions taken can never fully prepare one for the first step into a new world of unknown ventures.

Climbing down the side of the ship from the rope ladder that morning into the LCI, Robert reflected back on the training from the past few weeks. Yes, this was not new to him for he had accomplished this many times. This time, though, he was going into the face of his enemy. This time, though, he may not return. And, if he did not, would he be buried on this island half way around the world as he knew it. Those are factors impacting performance that no training can ever prepare anyone for. One must experience those things oneself, and come to an inner resolution. That resolve is to do the best you can, and the rest is totally up to God.

The LCI ride to the beach was silent. Hardy anyone spoke a word. Bleak searching eyes met fellow bleak searching eyes. A few prayers could be heard, but no one said anything out of disrespect. It is during those times which may be one's last moments in this life, that the soldier respects his comrade's belief in whatever they think will protect them, and/or carry them across the portal of this life into the next life. The LCI driver gave the order preparing to disembark, the LCI stopped and the ramp slammed down, and they all ran for their lives onto the rocky beachhead for cover.

The nine destroyers had furnished the landing assault with gunfire support. Because of bad weather, much of the landing movements were difficult ones. Nevertheless, naval gunfire and air support neutralized the small Japanese garrison at Saidor, and the landing assault was relatively unopposed. Even though the element of surprise had evidently been more complete than expected, the assault resulted in casualties; one KIA and several WIA.

The primary mission of capturing the airfield area was very quickly accomplished. Now a tremendous task faced the combat engineers; that of building docks and roads, and in getting the airfield into operation. Of additional strategic importance, the Saidor landing put American troops in a position for severing the Japanese land movement of supplies and troops along the trails between the coastal lines and the mountains. An estimated 5,000 enemy troops east of Saidor were now in danger of being cut off from their supply bases by both land and sea. As soon as the Saidor beachhead was established, U.S. patrols pushed inland for preventing movements between the Japanese bases farther west along the coast and the forces now trapped between Saidor and the Australians operating against them from the east.

On Sunday, 2 January 1944, the Allied assault hit the beaches of Saidor. That day became Robert's first combat baptism by fire; bullets away for a second Happy New Year day! Even though the assault went relatively unopposed, there was gunfire – especially U.S. Naval gunfire and air support. At the end of the second New Years Day, Robert lay in his tent thinking over the day's events. His thoughts were of the anxiety-filled morning, the sobering LCI ride, running for cover on the beach, patching up a few wounded men, and actually earning the title of 'Doc' for the first time. With his hand on his .45 automatic pistol he thought of his wife back home, was thankful of still being alive, and fell asleep.

Into The Jungles

At the Saidor beachhead, Robert ran into a Native American Indian by the name of Chief Clay. Chief Clay was at a tree in the tree line where Robert ran for cover. Combat introductions are always brief, and so was this one. However, when realizing this was Robert's first combat landing, and there would not be much enemy resistance, Clay instructed Robert to stick with him and he'd take care of him. Little did they know at the time, that brief introduction began a close friendship between those two GIs.

After taking Saidor, Robert went on many recon patrols into the inland jungles of the surrounding area, and sometimes Chief Clay would be in the patrols. Other times when not on patrols, the two men would talk about their lives back in the States. The Chief was an Oklahoma Cherokee, always carried a Thompson Automatic Rifle (Tommy Gun), and was usually the point man on their patrols. The Chief talked straight and so did Robert, which was probably why they 'clicked' so well as friends.

The patrols the men became involved with were for isolating the enemy and pushing them back to the sea if they did not surrender. Surrender, though, was not in the enemy's way of thinking. Suicide over capture was. Along the patrol pathways were seen many vivid illustrations of enemy atrocities committed against both the native islanders and the Allied forces. Also along the way, they very soon discovered an interesting fact about the New Guinea people – they hated the Japanese soldiers with a passion. Their hatred ran so strong they developed an unusual resistance group called 'Fuzzy Wuzzy Angels.' The Allies would often call the Fuzzy Wuzzy Angels just 'Fuzzy Wuzzies' in short. Their name was derived from the fact they had frizzy hair and were self-sacrificing. And, they did not render aid to any of the Japanese. Many times reports were known of Fuzzy Wuzzies intervening between the Allies and the Japanese in sacrificing their own lives so the Allies could live and go free.

The jungle, in general, became another enemy, and an insidious one. The lee side of the mountainous island spine was wet from January to April. On the windward side, (scenes for most of the ground fighting during 1942 to 1945), the annual rainfall runs as high as 300 inches per year. One GI reportedly stated, 'It rains daily for nine months and then the monsoon starts.' Because of the high annual rainfall, uniforms were continually wet and literally rotted on the soldier's backs. Blankets would become heavy with rain; then mold covered, fly-blown, and usually ended up being discarded. Men were always covered with festering sores – the jungle rot which plagued the GIs ever since landing on the island. Fevers and dysentery took their toll as well; malaria, dengue, and typhus took men out of combat as fast as the enemy's bullets at times. The medics were always handing out Atabrine tablets for helping fight malaria and dengue.

Yellow jaundice was another enemy the soldiers faced. Due to the environmental conditions inherent to the habitant, the affliction was not uncommon. Some stated that everyone over here had the 'New Guinea Tan' (yellow jaundice) in varying degrees at some point. Yellow jaundice, if gone untreated will shut down the liver and kidneys, and then you're done for good (dead). Yellow jaundice also sometimes came in the aftermath of malaria. Therefore, it was 'theorized' an increased intake of Atabrine for malaria was also beneficial in the fight against jaundice.

Fighting and killing in war was bad enough, but there were still the huge snakes, spiders, poisonous critters, and countless biting insects to contend with. Yet, elements of the 32nd Infantry Division proved victorious through all these things. They were very proud of their accomplishments since landing in the Solomon Islands in early 1942. So successful were they in jungle fighting, they boasted on 'having wrote the book on it.' After what the 32nd had endured in Buna and Port Moresby the year before, they had, indeed, earned the right to boast.

In spite of everything, Robert quickly adapted to his new job as one of the 'Docs' with the 32nd Infantry Division. He'd earned his title of Doc at Saidor, and earned titles went a long way with comrades – especially on combat patrols.

In one of the night patrols around Saidor, he was out with his assigned squad to recon a Jap encampment position. The mission was to verify its location, strength, supplies, and fortifications, and then report the intell back to their Intelligence Officer. The squad geared up before dusk and by nightfall began their trek into the jungle. The suspected enemy's position was several klicks away. Thus, by the time of reaching their destination, it was taps (bedtime) for the Japanese troops.

Their squad came to a clearing at the edge of the jungle, and on the other side of the clearing lay the enemy encampment. While stealthily reconnoitering the encampment as much as possible, one of the Japanese guards spotted their movement in the tree line. Calling out something in Japanese, the soldier raised his rifle toward them. When a properly secure response was not returned, the guard opened rife fire on them. They promptly turned and began running through the jungle back to home base. Their run, though, was not alone for very long.

Within what seemed like only seconds, the jungle behind them became alive with enemy shouting and gun fire. Bullets zipped past them as the GIs returned gun fire. As the Japs seemed to be closing the distance on them, they all decided it was high-time for running harder and faster and spending less time turning around to looking back and expending bullets. Bullets, though, have no eyes. When expulsed from a gun barrel, bullets travel expeditiously toward whatever they are pointed at. In this hit-and-run scenario, the Japanese bullets eventually found their marks.

By the time the squad had run about half way the distance back to base camp, Jap bullets had found the buttocks of every GI in the band-on-the-run, save one – Robert. 'How can this be possible?' Robert thought as they were all still running as fast as they could possibly go! Luckily, they'd been hit while nearing a downhill clearing point of the return run and the Japs ceased pursuit. The Japs may have thought the squad had a radio and called in for support. Since they were getting close to home base, the pursuit probably was a lost cause for the Japanese soldiers.

The band-on-the-run stopped just long enough for Robert to check their inflicted posterior damages. Luckily, they were not bad enough to cease their running, so they continued running for their lives. Once back at home base, Robert took the GIs to the field infirmary tent and with the help of a couple more Docs, began patching up the grazed buttocks of seven soldiers. During the patchwork, some of them sang fragments of the song, *The Battle of New Orleans* for making light of their painful situation; especially when coming to the words, "We ran though the briars and we ran through the brambles, and we ran through the bushes were the rabbits couldn't go..." All the while, through their aches, pains, and merriment, Robert heard those seven soldiers whine about why they got shot in the butt and how come he does not even get one scratch.

After enduring their friendly chides long enough, Robert paused during one phase of their shot-in-the-butt patchwork. He just smiled widely, pointed one hand upward, and replied, 'Well, fellas, I guess there must be one big Fuzzy Wuzzy Angel up there in Heaven somewhere who likes me.' And with that jesting comment, the unit's field medical tent in those predawn hours became ignited with even more laughter.

Aitape Landing

After being the sole buttocks survivor in a recon mission one night, Robert became quite popular with the GIs as the Doc of preference the guys called on to be in combat with because he most likely had a Guardian Angel accompanying him. Well, that may have been somewhat of a comfy feeling for them, but it was not so comfy for Robert in some respects. That meant he was probably going to take more chances than usual. It was the old law of supply and demand taken toward another risky dimension.

Realizing this concept caused him to carve more slowly on his wife's nickel wedding ring. Maybe that newfound purpose in living could be delayed long enough for getting him through this war and back home. However, he did want Pauline's ring home before their wedding anniversary in April if possible. In between jungle patrols and island hopping, he found the time to finish her ring and get it mailed home before the end of March. Hopefully it would arrive by their anniversary.

After mailing Pauline's ring, he began carving on his wedding ring. His carvings, though, were always interrupted with something. His carving began just as his unit was gearing up for an assault on Aitape. Before the assault, Robert had already grown tired of war. If he wasn't on jungle patrols, he was island hopping along the coastline of New Guinea. The Japanese still occupied many of those islands.

April marked the end of the wettest season in the Aitape region of New Guinea. In that area, rainfall averages about 100 inches per year. Although June is typically one of the driest months of the year, July could be one of the wettest months of the year. Thus, torrential tropical downpours, rather than prolonged rains, were an expectation for the assault on Aitape.

Japanese development in the area centered on an airfield construction near a large plantation about eight miles east-southeast of Aitape. Intelligence reports had indicated the Japanese ground defenses in the Aitape area were weak. Therefore, it seemed probable there would be little opposition to a landing; and the assault force, once ashore, could quickly seize the air strip area. As weather forecasts are many times unreliable, so also can be the military intell received. However, if the received intell is all the war planner has to go on for planning, then so be it and conduct the best plans possible.

15 April 1944 was initially planned as D-day for the assault on Aitape. However, it was rescheduled for a week later on 22 April. On 18 April, Persecution Task Force departed from the Finschhafen area of New Guinea, and was destined for Aitape. Then, in the early morning hours on 22 April, the assault on Aitape began as planned; and with heavy preliminary shelling and bombing of the beach and coastal installations. At 0645 hours on that day, the troop landings commenced as

scheduled. Under protection of naval gunfire and air support, all landing operations seemed to be going according to plans.

As previously mentioned with the Saidor Landing, training never fully prepares anyone for the real deal. Every assault landing, every skirmish, every fight is a brand new situation. All the details are totally different in each and every circumstance for the makings of war. All the repetitive actions taken can never adequately prepare one for the first step into any new venture.

As with many times before, climbing down the side of the ship from the rope ladder that day into the LCI, Robert began casting some reflections. This was not training, this was the real deal. He had survived the last major assault landing, but what about this time? He was resolved again to do the best he could, but the rest was totally up to God.

The LCI ride to the beach was silent, and hardy anyone spoke a word. Ghostly eyes met fellow searching eyes; looking deep into empty souls. Naval gunfire was heard streaming overhead and impacting the coastlines. Air support was also heard, and seen, bombing and strafing the beachhead. No one said anything, but as a few faint prayers were heard, most knew that everyone hoped the naval and air support would chop up the beach as much as possible before the LCIs made landfall. The time came for the LCI driver giving the prepare-to-disembark order, the LCI stopped, the ramp slammed down, and they all ran for their lives onto the sandy beachhead; trying to find cover.

The heavy naval artillery shelling and air support bombing had accomplished a good job of softening things up on the beach. Once in the tree lines, the safety zone from bullets seemed to come settling over as a blanket. Even though there was no such safety zone, the trees provided far more cover than the beachhead ever could. Robert had grown up in the tree covered hills of

Kentucky, but he never knew trees and tree stumps would ever feel as precious as when in combat.

Robert and the rest of his unit did not stay long on the beach, or the tree lines, though. As they began moving further into the jungles, the enemy was soon retreating! Tactical surprise by the battle planners was again achieved so that the troop landings had been accomplished without major opposition. The combat teams quickly established their lines after overcoming opposition on the beaches. Then, within two hours after initial troop landings, LSTs were unloading tanks on the beach. This then, enabled the GIs to press rapidly inland, and by nightfall the task force had established a secure beachhead, and had begun overtaking the airfield. The next day, following naval and air bombardments, amphibious units landed on Tumleo and Seleo Islands (just off the coast and opposite the beachhead). Seleo was quickly over taken but, on Tumleo, significant resistance was encountered.

These Allied landings at Aitape, Seleo, and Tumleo had virtually isolated an estimated 60,000 enemy troops. This successful assault on Aitape meant assuming more real estate. More real estate meant more jungle territory to comb the enemy out of. Since the Japs would not surrender, and since the jungle was also an enemy, a long and difficult task lay ahead for the invading Allied forces. Not to mention that during all this, Robert had a wedding ring with which his wanted taking his time carving.

Battle of Biak

Not long after taking Aitape, elements of Robert's unit were designated to become part of the Alamo Force Reserve for the assault on Biak Island further up the coastline. The Japanese had occupied Biak since early 1942. It was not until late 1943 that they decided on making the island a key air base which would be within fighting range for many of their other air bases in the western New Guinea area. Allied intell had discovered the Japanese had sent to Biak one of their best army regiments to protect and hold the island. It was a combat seasoned unit from China. Intell also reasoned it was probable the Japanese initially intended on establishing Biak as an important ground stronghold as well as a major airfield. Therefore, at some point the Allied forces *simply had to take this island*.

The role of the Alamo Force Reserve in the assault on Biak was to supply the major task force (Hurricane Task Force) with as much detailed information as possible after their landing. They were to penetrate the jungle immediately upon landing the beachhead, recon the deeper areas, and report on the enemy's capabilities. Even though aerial reconnaissance had exposed some intell, the further detailed extent of their defenses had remained unknown.

The dawn of 27 May 1944 came very early for Robert and some others. They had been up since 0200 in preparation for the first wave of beach landing. And, as plans go, they sometimes get changed – they were to go in on a later landing. By 0930 that day, they were running ashore from LCIs onto the beach of Biak, dodging pillbox gunfire and the strafing gunfire from Japanese aircraft. As the GIs took cover in the tree lines, they were hoping to not get hit by the enemy's mortar fire and bombers. However, while Robert was running, shrapnel from a nearby exploding mortar struck his medic pouch and ripped it open.

After taking cover, Robert discovered the shrapnel had damaged more than just his medic kit. It had also ripped open the fatigue pocket in which his ring was always carried. To his grappling surprise, the wedding ring he had been meticulously carving was gone! Taking a quick look around, and also trying to stay alive from strafing gun fire, he could not find the ring. Quickly residing on the fact that another ring could be made, but a life could not, he immediately moved toward more cover and higher ground.

During the following several days of fighting (and half of them were rainy), it was discovered that Biak was one of the worst possible places to put an encampment. The island had previously been an old fuel dump and the Allied bombing support had helped bomb it to pieces. The Japanese ended up retreating to some cliffs on the coast line, which were filled with caves. Because they would not surrender and kept on fighting, the GIs began blasting away with tanks, machine guns, grenades, mortars, and flame throwers. The battle sounded like an over-sized Fourth of July celebration day and night until it finally ended 27 days later.

Of the 12,000 Allied soldiers in the Battle of Biak, nearly 3,000 were casualties, of which 474 were KIA. Of the estimated 11,000 soldiers with the Empire of Japan, the total casualties were unknown. However, 6,100 were confirmed as KIA, and 450 were known taken as POWs. With these casualty statistics, the Battle of Biak had been another of the bloodiest ones in the Southwest Pacific during WWII.

During the battle, and while under gunfire, Robert came across a wounded GI and began rendering first aid. In the aiding process, a Jap soldier began shooting at them while charging their position. After expending his ammunition, the soldier kept charging while loudly screaming. Robert pulled out his .45 pistol and shot the soldier dead in his tracks as he collapsed a few feet from them. After patching up the wounded GI, Robert waited for some stretcher bearers to take the wounded soldier away. While waiting, Robert noticed the Jap soldier he'd shot had fallen across some other dead Japanese soldiers. After the wounded GI was removed, Robert pilfered the coat pockets of the soldier he'd killed. Some GIs in the war were extracting gold teeth from the enemy bodies, but all he wanted was a souvenir from the Empire of Japan. Hopefully, someday when back home in the States, he could look at it and say he'd been there, lived through the hell, and made it back home. Robert found a pocket-sized military field manual, a unit Rising Sun Flag, a rice bag, and a metal vial. Those were good enough souvenirs from the hell of those places.

After the clean-up process had begun on Biak, Robert and several other GIs were given some time to unwind. They decided to go swimming on the northeastern side of the island. About mid day, the incoming waves began getting stronger; so strong that they were catching the GIs and slamming them against the ocean floor. When surfacing, they would be nearly out of breath. One of the men saw a huge log floating further out in the ocean and began swimming toward it. The lieutenant in the group called to him stating it was probably too far out and it was not a good idea. The GI kept on swimming and, exhausted, finally reached the log. About that time, a shark's fin appeared further out in the water. Everyone closer to the beach swam as fast as they could toward shore. The lone GI was helplessly stranded on top of the huge log as it slowly drifted further out into the ocean – while being circled by the shark.

Once everyone was ashore, the lieutenant asked for a couple of volunteers to stay on the beach with the GIs belongings, set up a marker with some rocks and tree limbs, and then return to camp. The others made their way back to camp as quickly as possible to report the incident for a possible aerial drop of a life raft. By the time the incident was reported and a plane could be in the area, twilight was soon to set in. A patrol boat was dispatched to the beach where the markers were erected and then set out to sea for a search and rescue attempt. The GI was never found.

A few days later, the detachment of Robert's unit, which had supported the battle, was sent back to their home camp near Aitape. On the trip back, he endlessly inquired of everyone until he came across another New

Guinea nickel. It was when landing at Aitape that he actually acquired the nickel, and it was from a newly-assigned soldier to the unit. His name was Ryan and he was from Danville, Virginia. Interestingly, Private Ryan had been assigned to the 32nd back in the States, and had also departed Angel Island same as Robert. Ryan, though, departed Angel Island a few days earlier and landed at Midino. Ryan had been in another Infantry Regiment within the 32nd until the assault on Biak. Just prior to the Battle of Biak, he had been transferred to the 128th Infantry Regiment, which Robert was assigned to at the time.

When hearing of Robert's ring-carving project, Ryan thought it was a great idea. The ring project caused the two men to discover they were much alike and an immediate friendship was struck between the two GIs from 'the hills' back in the States. Thus, with the acquisition of another nickel, another ring project was begun. And, with that new acquisition, along with it came the sudden solemn thought striking Robert; what if the next time he took a hit, would it be a fatal one?

Once back at home base camp, they discovered their unit was relocating into another area – about twenty miles east near the Driniumor River. Allied intell sources indicated that over 20,000 Japanese Army troops were approaching the Driniumor River. Their goal apparently was breaking through Allied lines and retaking their loss of Aitape. Therefore, Robert's unit was among those sent to guard Aitape's Allied-occupied borders along the Driniumor River.

While they were making their way along a muddy road next to a small stream, Robert slipped, fell on some rocks, and sprained his ankle. Even though some other soldiers stopped to render aid, this set him back from advancing with the column. A couple of medics came along and relieved the other soldiers. They tied small tree limbs for splints around both sides of his ankle and helped him hobble along. It wasn't long afterward that

Robert heard a familiar voice behind him saying, 'Hey hillbilly, if Japs hit us, bein' all hobbled up like that, ya gonna get self killed.' As they turned around toward the voice behind them, they saw Chief Clay.

Chief Clay's big smile showed his white teeth shining brightly in the sun against the backdrop of his tanned red skin wrapped within olive drab. Robert said in reply, 'Hey Chief, this is one hillbilly ain't gettin' killed out here.'

'Oh yeah, and Great White Spirit tell ya that?' Clay said.

'Na, sometime ago on Saidor you said for me to stick with you and you'd take care of me. So then, if ya do your job, I'll be home safe, right Chief?' Robert responded.

'Well, I'll do my work here and rest is up to Great Spirit. It is He who decides such things' Clay muttered while looking around and with his Tommy Gun.

About that time, Ryan came walking up to the slower moving duo. Robert introduced the Chief to Ryan and after a few words of conversation, they seemed to hit it off as friends the same as he and the Chief had done. Movement for Robert was slower than planned, but by nightfall his unit had reached their new camp sight – he, Chief Clay, and Ryan trailing in last. Luckily for Robert and his unit, the enemy did not engage them in combat that day. Once setting up camp for the night, Robert treated his ankle sprain, and then he, Chief Clay, and Ryan began talking about their lives back home before turning in for the night. Early dawn came and along with it, so did something else – the sounds of loud snorting beneath them!

The three men had settled in for the night at the edge of the campsite. They had tied their sleeping bags between the trees and slept in them under mosquito netting as usual. A wild boar had wondered near, most likely out curiosity. Those animals were known to rip open a man's legs with their tusks and hamstring the muscles. Robert always slept with his .45 pistol, and therefore, he always felt somewhat 'protected.' However, he did not want to fire off any rounds in the quiet dawn hours and set off false alarms to the guards on watch. About that time the other two men awakened and eyed the wild boar.

Semi-frozen in their cots, the three men watched a potentially good pig roast meander around the campsite snorting and sniffing as it cautiously groped along. A few moments into the walking-pig-roast scenario, one of the guards on watch came around the edge of a tent and spotted the wild boar about the time it saw the guard. The wild boar turned scampering off into the jungle, and so did Chief Clay, Ryan, and a couple other GIs who had also been observing the 'swinesome' sight. Only a few moments later, some gunshots were heard, soon followed by one of the men returning into the camp looking for a Jeep and some rope. The wild boar had been killed and they needed the rope and the Jeep to haul it back in to camp.

The men at that campsite had a most welcome change in their dining menu that night besides standard K-rations – wild pig roast.

Battle at Driniumor River

Even though Robert's first baptism by fire came at Saidor, many other skirmishes, attacks, and assaults honed him for enemy fighting while patching up the men in his unit. None of those things, though, prepared him for what was to become known as the Battle at Driniumor River. Within any U.S. Army Division, personnel can be transferred alternately between its subordinate units. At the time of the Battle at Driniumor River, Robert was assigned to the 128th Infantry Regiment. His duty assignment at that time became one which would etch the horrors of war in his mind long beyond this battle and this plot of land.

Although the Driniumor River varied in width from around 75 to 150 yards in the areas where the enemy was expected to cross (according to intell), it did not present much of an intimidating obstacle. In those areas the river was usually only about eighteen inches deep (except after heavy rains). Thick jungle vegetation on both banks made it difficult in detecting movement on either side. Because of the large enemy troop buildup, Allied field planners deemed it necessary for establishing a delaying position on the west side of the river.

The planned delaying position mainly consisted of two infantry units within the 32nd Infantry Division; the 127th and 128th Regiments. There were others, but these two units were the main Allied defense force for the holding line. If the Japanese attacked, these soldiers in the delay position were to hold this line as long as they could. Then, if necessary, they were to withdraw westward and establish successive delay positions until they were pushed back to the main defensive position around the airfield.

For days, the Japanese shelled and fired on the infantry units in the delay positions. As the Allied intell gradually increased, unfortunately their unit strength was not increased to the strength level with that of the enemy. These units continued patrolling along their side of the Driniumor in an effort to keep track of the enemy's disposition. As they did, they met increasingly more opposition. Thus, it was becoming increasingly evident to higher command that the Japanese were apparently preparing for a significant assault in retaliation against losing Aitape.

Early on 10 July a recon group was sent across the Driniumor to conduct some reconnaissance, and they met significant enemy opposition. Aided by good artillery support they obtained intell, but suffered 5 KIA and 8 WIA during the day. The delay position was now responsible to hold a front of 3,000 yards on the right end of the line. The 127th was spread out for 1.5 miles in the center sector. The 128th was responsible for the remainder of the line on the left, which stretched nearly 3.5 miles to the coastline. Patrols were sent out all along the line throughout the entire day, and all of them made contact with the Japanese. The entire front line of five miles long was told to be prepared for an assault that night while the units conducting the reconnaissance were ordered to continue pushing eastward the next morning.

The Japanese were caught by surprise at Saidor and Aitape, and very soon they would be repaying it.

Shortly before midnight of 10 July 1944, after a short artillery preparation (which came as a surprise because no enemy artillery had been identified within range of the Driniumor River), approximately 10,000 Japanese infantry troops began charging across the river. They came in suicidal waves against the front line sector primarily held by the 128th Infantry Regiment, while loudly screaming 'Banzai! Banzai! Banzai!' This was known as a Banzai Charge, was suicidal, and was deemed an honorable death by Japanese soldiers. As the Banzai Charge ensued into the night, Japanese bodies had piled up in front of the machine gun embankments so high that the machine gunners could not fire over them. The machine gunners were forced to leave their fox holes and pull bodies from the line of fire, just so a clear line of fire could be sighted on the enemy. This scenario was repeated during the numerous Banzai Charges throughout the course of the night.

The enemy came over the river like waves of ants while the American GIs poured out machine gun and automatic rifle fire until their barrels turned red hot. The Japanese, even though dimly visible under the light of the ignited flares, continued rushing headlong on toward their deaths. Allied artillery fire fell in clusters onto the Japanese soldiers, killing and wounding them by the dozens. Fox holes became replaying scenes of mini-battles with GIs and Japs locked in desperate fighting to avoid the inevitable bitter end of death. Every GI, regardless of his assigned position, literally fought for his life and that of his comrades – by gunfire, grenades, and bayonets in hand-to-hand combat for the entire night.

Many of the enemy also became crushed beneath the falling palm trees that were being snapped apart from the artillery explosions. Despite suffering staggering casualties from machine gun and artillery fire, the Japanese pressed on forcing a breach in the American front line. When morning light finally arrived after what seemed an eternity, the dawn exposed a grim reaper's reality; soldier bodies by the thousands were scattered everywhere. In some places, they were piled high enough for making human mounds.

That one night of battle at the Driniumor River had turned into a Grim Reaper's carnage beyond anyone's comprehension. Brave GIs who had fought for their lives all through the night, stood in the dawn's light at the horrendous spectacle around them. They hung their heads, and wept; crying for their buddies, for themselves, and for all the slaughter beneath their feet.

At the point where the Japanese had breached the line, the GIs had run out of ammunition – and had been fighting in hand-to-hand combat. Under provisional orders to withdraw for establishing another delay position, and after having gone through a traumatic night of fighting, those GIs withdrew into the jungle. For the next three days they managed, wherever possible in the jungle, to engage the enemy with gunfire and hand-to-hand combat. By 13 August, the battle-weary GIs began counterattacking for sealing the breach in the line. In sealing the breach, they obtained valuable fire support from U.S. and Australian aircraft, from two U.S. destroyers, and from two Australian cruisers.

After the breach was sealed, the enemy was still behind the Allied lines in isolated pockets. The succeeding several days were filled with many sneak attacks from Japanese snipers. As the battle raged on from days into weeks, the Japs would utter cries such as 'Help', 'Medic', or 'I'm dying' in attempts to lure the field medics and other GIs from out of their foxholes so they could shoot them.

It was during the time attempts were being made for sealing the breach that Robert jumped into a foxhole and began rendering first aid to a wounded GI who'd called for a medic. While working away, strafing fire came across the foxhole and killed the GI he was attempting saving. Ducking further down into the foxhole for cover, strafing fire came back across again where he had just been rendering aid. He picked up the dead GI's rifle and began spotting the Jap sniper for returning fire. Every time Robert peeked up from the foxhole's edge, strafing fire rippled across it. As the minutes seemly began slowing down for a long waiting period, he came to a shocking realization. He was pinned down and the Jap's strafing fire was continually zeroing in closer.

About the time Robert thought the end for him was near in the Jap's gun sight, Chief Clay popped from around behind a nearby palm tree, and ripped up the side of another tree with his Tommy Gun until his bullets found the Jap sniper. A loud scream announced death had found its mark and the Jap fell part way out of the tree – he had tethered himself to its branches with a rope.

'The Great Spirit wishes you to live, fix us, and fight more days' came the words beaming from Chief Clay's wide smile.

'Yeah, and I see He chose the lead from your gun as medicine for me this time, too' came Robert's reply as he crawled from the foxhole.

The next day, Robert, Ryan, and Chief Clay were on a patrol a little further up the river. Their squad had come to an opening along the bank and was taking a break. One of the GIs was at the bank's edge acting as their lookout, some of the others opened some K-rations, and some others began obtaining water for their canteens at the river's rocky edge. After a few moments, the lookout spoke up about possibly seeing a log jam floating down the river. Everyone got up and came near the lookout. As the perceived log jam began coming into view, one of the soldiers spoke up, 'Heyyyyy guys, them ain't logs......them's dead bodies.'

As the assumed log jam edged closer and its vague outlines became more refined into human forms they definitely were dead bodies – mostly Japanese soldiers. The water became tinted red with blood as the gory sight slowly began passing by them in review. Everyone who'd opened K-rations ceased their eating. Those who'd filled their canteens empted them and slung them as dry as possible. After only a few minutes of this gruesome display of death, the squad quietly gathered up their gear and moved on with their recon detail.

About a week later, and after many skirmishes had been encountered, Robert and Ryan were on patrol with their squad. They came across the scene of an intense skirmish between several GIs and the enemy. In looking over the casualties, taking dog tags, and trying to cover up the bodies until a Jeep could return for them, Robert came across a headless body he recognized – that of Chief Clay. His head had been severed by sword in hand-to-hand combat, and his upper torso had sustained wounds from an explosion. Robert just stood there in shock, for his closest friend in this senseless no-man's land, was now forever gone. As he took the Chief's dog tag, he also took his tribal necklace. Even though the Chief was gone, Robert began carrying that necklace everywhere as token of a once great friendship.

The remainder of July saw heavy fighting west of the Driniumor as various-sized patrol units clashed with the enemy in the jungle. Scattered all through the jungle around the mouth of the Driniumor were pockets of

American GIs still holding their positions even though they became encircled by the Japanese troops who were determined on wiping out the Allied forces. By the beginning of August, however, the Japanese were forced back across the Driniumor River. The U.S. troops continued their annihilation of the Japanese forces as the enemy retreated further east until the battle was officially declared over on 25 August. The Battle of Driniumor River officially lasted forty-five days, but the main intense fighting lasted thirty days. It became the second costliest campaign in Papua, New Guinea.

In the Battle of Driniumor River, the Americans suffered almost 3,000 casualties including 440 KIA. The Empire of Japan, however, suffered nearly 10,000 KIA. Those who survived the hardship and terror of that horrendous battle are the only ones capable of providing any insight into what really occurred during those intense days and nights. Their battle was a savage one fought in primitive conditions. Their battle was laced with mass suicidal intentions on one side, but it was outmatched with a will to live and be free on the other side.

When the fighting began to subside after 10 August, the GIs finally found some time for cleaning themselves up. Most had not bathed in nearly thirty days. Robert and Ryan finally sat down in a secure area to take their boots off for the first time in a month. Their boots were tattered and almost rotted into pieces. When they did come off, their socks were gone. Their socks had completely rotted from off their feet. The sores that replaced them signaled jungle rot was not far behind.

The average mind cannot comprehend over 10,000 dead bodies within a five- to six-mile radius. Actually, it was only half of that area because the area was a beachhead with the terrain leading up river toward the mountains. Robert became one of the soldiers his unit designated for body detail in the aftermath of the battle. For days he loaded bodies onto trucks, carried bodies onto the beachhead, and tagged them for transport. There was one time he saw stacked bodies high and in a line on the beach for as far as he could see.

In the past, Robert had always used his Lucky Strike cigarettes for nose filters when having to be around decomposing bodies. It was during that bloody body detail after the Battle of Driniumor, though, that Robert found the scent permeated through the cigarettes more than any of the other times. The scent of death also pierced more than his sense of smell – it penetrated his soul. He never forgot the experience of that detail, and he requested of his superiors to never be placed on such a huge one ever again, and he wasn't.

Someone once said that life stains the soul. Thus, it would be the good *and* the bad of life that determines how *deep* the *stain* and how *intense* the *aroma*. It is the stench of etched-in stains on the soul we so much wish to wash away the longer we live. The deep stains of war began etching into Robert's soul at the Battle of Driniumor River, and they would surface in his nightmarish dreams for many years to come.

Island and Jungle Patrols

After the gruesome body detail at Driniumor, Robert was transferred up the coast to Hollandia. Rumors were that the 32nd Division was being readied for General MacArthur's push on taking back the Philippine Islands sometime in the fall. Once settling in at Hollandia, they continued jungle patrols from there and also did more island hopping. As before, for some island patrols the GIs were assigned, they were only given a number which represented the island. This practice was for security reasons. If the GIs were captured, the less intell they knew, the less information the Japanese could extract from them.

The squad Robert and Ryan were out on patrol with one day had landed on an unknown island for an intell recon. There were some suspected enemy encampments on one side of the island, and their mission was observing the enemy's strength, supplies, and artillery fortifications. As they were reconnoitering the encampment from some trees at the edge of an open field, they saw some natives conducting some sort of tribal dance. While their photographer busily snapped away a few photos, one of the Jap guards must have seen the camera lens reflect some sunlight for he raised his rifle and pointed it toward them.

It was then some Fuzzy Wuzzy Angels came from the native's ranks and began dancing in the line of sight between the guard and Robert's squad. That intercepting action given by the Fuzzy Wuzzies was all the heads-up the squad leader needed for the GIs immediate departure. As they quietly turned to leave, they heard the rifle fire, the natives roar, and then more gunfire. The GIs never stopped to look back, but leaped into running toward the coastline for their pick-up. Had it not been for those Fuzzy Wuzzies, Robert's squad (or some of them) may not have come back to home base camp that day.

Fighting skirmishes would always flare up somewhere on those miserable islands. Seemingly they would occur more often now that rumors of invading Leyte Island were around. However, rumor control had it that a day in mid October would be the D-day for the Leyte invasion. Then rumor control of the invasion became more serious when discovering elements of the 32nd were being relocated to Morotai Island in mid September. Plans supposedly were, once the island had been secured, it could be used as a staging theater for an invasion of Leyte Island in the Philippines.

15 September came, and Allied Forces, along with elements of the 32nd Division, landed on Morotai Island. Even though enemy resistance was not as strong as initially expected, resistance was met; and fighting on the island was continually engaged until the end of the war. Then October arrived with a surprise. One day

while working on the finishing touch for completion of his wedding ring, the flap to Robert's tent came open and Ryan's head popped in saying, 'Hey, hillbilly, are ya ever gonna finish whittlin' on that ring?

Robert replied, 'Well, if I'd had that Indian Chief along with me at Biak, guess I wouldn't have ta be whittlin' on this new one.'

With that introduction, the two men sat for a while and caught each other up on the missing days of their wartime lives. Ryan had transferred in from Driniumor a few days before, was settling in, and asked where Robert's tent was. He brought with him more rumors of their unit going to fight in Leyte, and very soon. While hearing this news and talking of other things as well, the ring was finally finished; and Robert proudly slipped it on his wedding ring finger.

A few days later their unit began receiving briefings on their part of the Leyte Invasion. So, this time rumor control had changed to reality. When they weren't chasing after the enemy, they were practicing charging them when they hit the beaches of Leyte.

Fighting at Leyte

Mid October arrived and so did other unwelcome things for Robert, the beginnings of a tinting in the color in his skin – and it wasn't a tan. He had been suspecting the possibility of yellow jaundice, but was not positively sure. With this aliment, the whites of the eyes and the skin turn yellow. As the condition advances, the urine will darken. When this occurs, the illness arrival is official. In its advanced stages the liver and kidneys shut down and death is the final result. What a time to suspect such an illness – just prior to a major offensive invasion of an island. This concerned Robert for he knew there would be occasional weak spells. This was a further concern as soldiers needed to be at full physical operational capacity for increasing their chances of survival in combat. In considering all these things,

Robert elected for taking a path of self treatment – the intake of as many Atabrine pills as permissible and flushing his system with as much water as possible.

Robert continued that course of treatment until he found himself on an army troop transport ship destined for landing on Leyte Island just after the initial assault waves landed on Leyte the morning of 20 October. While sailing to the Philippines, some transport ships met trouble. Japanese planes attacked a few of them, but there were no losses. Such was the same with the voyage of the ship Robert was aboard as it drew near the Philippines. As they came in sight of Japanese patrols, one of the times when sighted by the enemy, they came under aircraft gunfire.

The Japanese were well aware of General MacArthur's intention for retaking control of the Philippine Islands. They were fully aware of their mounting continued defeats in the Pacific Theater of War. And, they were also fully aware of General MacArthur's broadcasted statement from Leyte beach on 20 October – 'I have returned.' Therefore, the Japanese soldiers were frenziedly suicidal in their resolve for NOT surrendering any grounds to the Allies. They were intent on fighting to their death if need be.

All throughout the New Guinea and surrounding miscellaneous islands, soldiers could go to sleep in a dry foxhole. Then, when the rains came in during the night, by morning they would be bathing in a running-over muddy foxhole. Things were no different in Leyte. The rains, mud, insects, snakes, disease, and the jungle, in general, became their enemies in Leyte as well as the Japanese soldiers.

After landing by LCI on the western beaches, Robert's unit soon went into combat action along the Pinamopoan-Ormoc Highway, taking Limon and smashing the Yamashita line through harsh hand-to-hand combat. The line was named after Japanese General Yamashita and was a heavily fortified area of Leyte to where the Japanese had retreated after sustained bombing by the Allies. As the intense fighting wore on, the GIs struggled through rain, mud, jungles, mortar fire, and artillery fire; while engaging the enemy and eliminating their fighting positions. It was during an especially fatiguing and intense fighting episode one afternoon that Robert responded to a soldier's call of 'Medic!' Coming near the foxhole of the blood soaked GI, the muddied face of Ryan looked up from underneath its dented helmet.

'Hey, hillbilly,' said Ryan, 'need a quick fix-up job so I can go kill some more Japs'.

'Well, hillbilly, let's see what the Doc can do for ya. Hold still and take it easy, okay?' responded Robert as he quickly looked over the GI's wounds.

The initial battlefield assessment was not a good one; Ryan's wounds were so bad he'd only have a little while longer to live. Robert applied some bandages to stop the profuse bleeding, gave him a shot of morphine, and talked to Ryan for a few minutes until the Angel of Death came and extinguished the light in his eyes. Robert just slumped down into the foxhole and wept. He only had two buddies in this war that he felt really close to; it had taken one, and now it had taken the other one also. Was it going to kill him next as well? Battle fatigue, losing buddies, missing home, and wondering if it would ever be seen again will suck the life out of any GI.

Ryan was a soldier who believed in God, and always carried a Bible with him. Robert removed one of Ryan's dog tags, and then reached into Ryan's ammo pouch where he always kept his Bible. Removing the Bible, he placed it on Ryan's chest and whispered, 'I'll take care of these and get'em home if I can.' A few seconds later strafing fire rippled across the foxhole and Robert had to quickly duck down into it. When locating the enemy, and using Ryan's rifle, he shot the charging Japanese soldier and moved on through the hail of gunfire toward another agonizing cry for 'Medic.'

All during the early days of Robert's involvement with the assault on Leyte, he was not feeling well. Slight chills and fever led into nausea, which he self treated with the GI's wonder drug of Atabrine. Then one evening when his urine began more discoloration, he wondered if this war was actually going to take him out as well. Maybe he could make it just a few more days.

Those few more days were all Robert got. While engaged in action near the Breakneck-Kilay Ridge area of Leyte, Robert came down with malaria and became extremely sick. Malaria, in addition to the onset of yellow jaundice, can become fatal very quickly if not properly treated. His fellow medics immediately had him sent to a field medical station on the coastline. By the time of his arrival, the whites of his eyes as well as his skin had turned very yellow, his urine had turned very dark, and his fever went high. The field doctors immediately transferred Robert to the hospital ship USS *Mercy* (AH-8) which was operating in the Leyte Gulf.

Robert remained on the USS *Mercy* for several days undergoing treatment for recovery while the medical staff ran diagnostic testing. Testing and diagnostic results revealed the illness had gone into hepatitis, he was not going to recover anytime soon, and he would also not be strong enough for combat duty anytime in the near future. Sometime during those recovery days on the hospital ship, he encountered something very strange – a vision. He was not delirious, for in delirium the images are not defined; whereas, in visions, the images are. In delirium, any imparted messages can seem somewhat vague; whereas, in visions, messages are clearly recalled. In this vision, he was on a battlefield and death had come to himself and many others all around him. After the vision, an uncanny sensation settled over him that, if he returned to the Leyte battlefields, he would certainly die. His letters home to his wife mentioned of this vision just in case something would happen to him when returning to battle.

Robert was not afraid, for he was the caliber of individual to never back down from anything. He accepted obstacles as challenges. He assumed that God was giving him a heads up that if he was ordered back to Leyte the end of his life would come. He, therefore, should begin preparing for it. Such would be the proper course of action the reasonable person would set into motion when given an advance notice from the Portals of Heaven. Maybe, just maybe, all his prayers to this God he did not understand actually did some good after all. Maybe God really did love him for who he was, and wanted him to be ready when death came.

Now that, indeed, was a very loving God whom Robert would have to think over and over in the coming days.

Once all the medical testing results were gone over the final time by the attending physicians, their recommendation was for Robert being transferred back to the USA for a period of rest and specialized recovery. By the first days in December, Robert was transferred to a transport ship and was on his way across the Pacific Ocean and back to America. His involvement with this war was done, or so he thought…

Raging Seas

Those personnel orders showing Robert bound for America sure had a warm and welcome appearance about them. Only about 6,000 more miles and maybe twenty-three more days and these pockmarked, mire, and mud-filled hellhole islands would be behind him. Even though the transport ship was moving, it conveyed a unique secure sensation all its own more so than all those months of rain, mud, insects, snakes, rotten clothes, kunai grass, and some of the local cooking. The kunai grass at times would grow over ten feet tall and could cause GIs to feel as though they were the prey roaming in the wild grass. Cooking was another thing in the islands; some was okay, but other items were sickening. Cooking fruits and vegetables with animal blood was just not right. One food scent he'd not be missing was the smell – monkey pot roast. In his mind, that was a most disgusting meal and he never wanted to see another monkey for the rest of his life.

The worst ingredient of the entire war that permeated his days was the smell of death. It was everywhere, and it was especially during the days following battles when on detail for cleaning up its aftermath. That stench carried the undeniable signature of man's inhumanity to man. He never wanted to come near that scent in such intensity ever again.

When the ship got underway, Robert was not topside to watch its departure from the docks. He was bedfast in the ship's sick bay, too sick and weak. Malaria and yellow jaundice had hit him hard and had severely weakened him. When entering the military, he weighed 142 pounds; and during boot camp and advanced training, he put on a few pounds of solid muscle. Now his weight was down to 133 pounds. Combat fatigue, dysentery, jungle rot, malaria, and yellow jaundice had taken its toll on his frame, especially in the past few weeks. Hopefully, the stateside voyage would add some weight to him before reaching its beloved shores.

After a couple of days out at sea, Robert began feeling some better, but he never got over the sea sickness at all. The voyage to Australia over a year prior, and all the other ship journeys, never bothered him, but this time it did. The fresh sea air and sunny blue skies seemed to have healing balms in them, but never quite enough for he always found himself hurling stomach contents over the side rails.

About a week out, the seas started becoming choppy and the skies turned dark gray. Within less than twenty-four hours the ship was being tossed in tremendous waves. By morning of the second choppy day, the waves were so high and strong, the ship began taking on water. Robert became so sea sick he hurled up everything in his system and then some. By noon, the U.S. Army Captain in charge of his transportation party made an announcement. He informed all of them the ship was in danger of sinking because it was listing too much at times and was taking on too much water. The captain ordered them all to don their life jackets and standby for 'Abandon Ship' when given the order.

By late afternoon, the order was given for all hands reporting to their stations for standby of the order of abandoning ship. Robert ignored the order. He was so sea sick he could not stand up at all. As the GIs in his transportation party filed by his rack, they all tried persuading him in coming topside with them, but, he said no. They even said they would carry him, but, he still said no. He was so sick he could not stand things anymore.

After the GIs filed out of their compartment, the captain came and asked Robert what kind of stunt he was pulling. Robert just responded of not pulling any stunt, and that he was too sick to go anywhere. The captain then gave Robert a direct order to get up, go topside, and await orders for abandoning ship. Robert just looked at the captain and said, 'No Sir, I'm so weak with sea sickness I can't even stand up. I'm tired of it all anymore and I'm goin' down with the ship.'

The captain just looked at Robert and said, 'Ok private, have it your way. Your life's your business. But I've got others to save as many as I can. Best of luck to ya soldier.' Then he turned and disappeared though the compartment door.

While the ship violently tossed in every direction, Robert's thoughts tossed him in all directions as well. He had lived through hell in those islands. He had seen blood and entrails mixed with sand and smeared over everything. He had seen sights in which those atrocities were forever etched in graphic images within his mind.

He had been fighting alongside men dying in the rain and mud, and of diseases. One best friend saved his life, but he could not return the favor. Another best friend died while he was rendering aid to him during his last few moments in a foxhole. Robert came down ill himself, was evacuated from the battlefield, and was too weak and sick to make it off a sinking ship somewhere in the Pacific Ocean on the way home. Such poetic justice for doing the best one can and serving the Red, White, and Blue!

Robert thought about all these things as the transient compartment furnishings were being rearranged by the raging seas, and him about to go along with them. The furious seas without, and the rampant storm within, brought to bear the loneliest hour of his life he'd ever felt. He did not want to face this time alone. Looking through a porthole at the raging seas, Robert called out to God and said, 'God, I've read where You can do all things. I've also read where You have calmed the sea for Your disciples. I'm not sure if You can hear me because I'm unsure about who You really are. But God, if You can, calm this storm down enough so we can get through it, and get home safe – oh, and also that we all can get home on time if possible! Thank you and Amen!'

The raging seas did not suddenly become calm as did the recorded events in the Bible. The raging storm within Robert did, though. It is when men come to solid grips with their infallible God, that He will then control their unstable destiny. Amazingly, within about an hour the raging seas became somewhat calmer – calm enough the preparatory abandon ship order was reduced to an advisory. As the hours ebbed away, so did the threat of the ship sinking. As the next couple of days passed, the skies cleared and the sailing became smooth. Robert's strength improved enough for him to go topside again. Still, though, he suffered from slight sea sickness and would hurl over the side rails. He was slightly sick the reminder of the voyage until the ship landed at the Port of San Francisco, California on 25 December 1944 – Christmas Day.

After stepping his feet on the docks of San Francisco, and when finding a clear place to kneel, Robert knelt and kissed the dock. He had made it back to America! He had made it back home! Kneeling and kissing the ground may have seemed strange, but he did not care. He was not the first to kiss the ground, and others followed with the same actions. One does not fully comprehend the emotions of a returning soldier to his homeland unless they have been in the same situation. When one risks their life seemingly countless times in foreign lands, among foreign gods, lives through death, destruction, and hell on earth, and then finally comes home to their loved ones – their homeland is most precious. This concept is especially true for American GIs. They are soldiers of *the red, white, and blue*; for they leave their footprints of freedom around the world.

Robert's travels found him back in Fort Ord again for hospitalization and treatment. After a few days he was transferred to a hospital in Jackson, Mississippi, for further treatment specializing in combined post malaria and yellow jaundice complications. On 15 February 1944, Robert received an Honorable Discharge from the U.S. Army with a thirty percent disability. He may have been discharged from the army, but the raging seas of war still lingered within. Scenes of life ravaged by the death of war would continue flashing within his mind. The haunting nightmares that war imparts as its companion would not leave, but would diminish over time. Robert may have made it through his part of the war, but the war left its indelible imprint on his soul for the rest of his life.

Killing and being among the killings may come easy after a time, but living with the killings over a period of time is the hardest part of all. And so it is with weary, combat-worn veterans; the indelible scars which are carried within are those wounds that Purple Heart Metals will never show.

Finally Made It Home

As a patient in hospitals at Fort Ord and in Jackson, Mississippi, and in transit between those facilities, Robert kept in tune with radio broadcasts and gleaned the newspapers. He eagerly listened for reports of any news the media could provide about what was happening with ongoing actions of the 32nd in the Philippines. Before being discharged from the army and returning home, he discovered the 32nd Division, with elements of the 1st Cavalry Division, entered the area of Lonoy, Leyte on 22 December. The combined fighting strength of those two units brought an end of enemy resistance in the upper Ormoc Valley. Then, from Leyte, the 32nd Division moved to the Lingayen Gulf of Luzon on 27 January 1945, and had pushed up the Villa Verde Trail on 30 January.

It would not be until months after Robert returned home that he would discover the heavily-suffered casualties on Leyte in December of 1944; the month he was in transit back to the USA. He also discovered the 32nd had engaged in more than one hundred days of bitter fighting from the Villa Verde Trail until they finally took Imugan. In all of those days of fighting, he discovered many of his buddies had been KIA. He also discovered the 32nd Infantry Division Medical Detachment had a field medical unit along the Villa Verde Trail. It had taken a direct hit from Japanese artillery fire. Many of the medical unit's wounded soldiers were killed in that hit; some included the medics – those whom he knew. Robert also discovered that along the Villa Verde Trail there were areas in which Japanese snipers were especially shooting the medics and stretcher bearers and easily 'picking them off.' Apparently, during that campaign, the 32nd Division had lost many medics, and Robert had known most of them.

On the trip home, and for many months after returning home, Robert would ponder the vision he had on the hospital ship of being killed in combat. He would play that vision over and over in his mind, and question all the whys about it. In addition, he would endlessly wonder why he survived the terrors of combat while so many other GIs did not. He found, though, there actually is no answer to that last question. Over time he just learned to thank God for the life and freedom resulting from all the hell and sacrifice he had gone through.

Then one night in the late February to early March of 1945, in an old farm home nestled in the hills of Eastern Kentucky, Pauline Adams and some of her siblings were asleep in the attic bedroom of their parent's homestead. Her younger brother, Richard Thompson, was awake, wondering about what life's pathways had in store for him in the years ahead. Lying there in his thoughts, enveloped within the security of home and surrounded by some of his brothers and sisters, he drifted back off to sleep.

In the early-morning hours, Richard was awakened to the sound of hearing some noises downstairs. It was the familiar sounds of someone passing in the night hours stopping by, needing to talk with his parents. These things would sometimes happen in those Kentucky hills where friends were always welcome anytime. However, something was different about the sounds from this nighttime visitor. His voice seemed very familiar. Then as Richard was looking toward the dim light glowing from the attic floor opening, he could see the lamplight shadows foretelling of someone at the base of the staircase.

With sleep foregone, his widened eyes saw the shadows announcing footsteps coming up the stairs. His eyes squinted into the piercing darkness as those shadows suddenly molded into the head and shoulders of a man entering the attic opening. Then came an army soldier quietly stepping onto the attic threshold and immediately came cries that Robert had come home from the war.

Lying on a quilt and mattress near his sister (Pauline), Richard peered ever so inquisitively around an old clothing trunk. He watched as the soldier walked over to where Pauline lay asleep. The soldier stopped, paused, and then fell to his knees before his wife. With both his hands and arms thrusting beneath her, he scooped her up and toward him. Embracing her ever so tightly into his chest and with one hand rubbing the top of Richard's head, Robert cried out, 'I finally made it home!'

Private Robert Adams had, indeed, made it home. He had gone from glory days through some bitter hell holes of war to get back home, but he had finally made it home!

Glory Days Epilogue

In this novelette story of *'Glory Days'*, the Robert Adams character in the story was my father. All that was written was true to the best of my knowledge. At a family reunion about three years before writing and the publishing of this book, I heard, for the first time, the story of my dad coming home from the war to the hills of Kentucky. My Uncle Richard Thompson shared those details with me. He and I had been talking about my parents, and he began reflecting back on some of their good days of years long past. As tears began welling up in his eyes, he told me of the night seeing my father come home from WWII. Through those tears he said it was a sight never to be forgotten.

Hearing of the story from my Uncle Richard prompted me into pondering about it on the way home that evening. For the next several days, elements of the story became interwoven throughout the avenues of my mind. Those elements prompted me to write the story

'*Finally Made It Home*' in two of my other books titled *God*Prints* and *DreamWeaver Storybook*. Those writings and ponderings further created a desire in me for writing another book someday; a book of military stories which would become this book.

As a small boy, I learned that my father was a U.S. Army Combat Medic during WWII and had fought in the Solomon Islands of the South Pacific. Growing up like any young boy, I was eager to hear his war stories. Those stories, though, were seldom heard. In fact, during the early days of my boyhood life, my father almost never talked of his combat days. Most all he ever said to me about them was, 'Son, I hope you never see what I have seen.' What I did hear was gleaned from listening to him and some of his friends and relatives tell of their wartime stories, which was very rare. Most of what I heard was, indeed, very bad. My mother told me that my father came home from the war a very quiet man, and remained so for a long time. For years, he suffered quietly from what was known back then as combat fatigue. Today it is known as combat stress reaction (CSR), or sometimes called shell shock. Post Traumatic Stress Disorder (PTSD) also can be associated with combat fatigue. Both CSR and PTSD can appear in varying degrees, and it can be buried within the layers of one's psyche, only to resurface years later – and sometimes reappear with problematic complications which may require professional medical intervention.

I still have memories, while as a young boy, of hearing my father cry out in the middle of the night and scream, 'They're coming, they're gonna get us, they're gonna kill me. No, no, no!' Many times in the midnight hours, I would hear my mother tell my father to wake up, he was not back in the war, that he was safe at home; and I'd sometimes hear my father breathe a sigh of relief and then cry.

My father suffered quietly through another war for many years – one of inner survival. He wondered for years about the vision coming to him one night in the hospital ship. Why would God let him know his life was coming to an end when going back to combat? Why would the Creator of all things do such? If God knows all things, then He surely knew the doctors were sending him back to the USA. Why then the vision? This vision would puzzle him about God for years.

My father had difficulty seeing it was the gracious loving hand of God giving him a 'heads up' that if he went back into battle, the conflict would not go well for him and those near him. His time on earth would be finished, and he should make things right between himself and God. In later years my father would make profound personal discovery; that God is a loving God and always wishes for people to get their hearts right before leaving this life. References for those concepts are found in Ezekiel 33:11; *"...I have no pleasure in the death of the wicked, but that the wicked turn from his way and live. Turn, turn from your evil ways! For why should you die..."* and also in John 3:16; *"For God so loved the world that He gave His only begotten Son, that whoever believes in Him should not perish but have everlasting life."*

In my researching data for this book, I discovered something quite interesting. My father had related to me several times about the ship nearly sinking on his return voyage back to America. At that time in December 1944, he was totally unaware of Typhoon Cobra. In writing this book, my research revealed that Typhoon Cobra struck the Philippine Sea at its ***highest impact*** on 17 December 1944. Typhoon Cobra was also known as the Typhoon of 1944. It was also known as Halsey's Typhoon (named after Admiral William Halsey) and was the U.S. Navy's designation for the typhoon which struck the U.S. Pacific 3rd Fleet on 17 December 1944.

Despite the given warning signs, on 17 December Admiral Halsey had inadvertently sailed his Fleet (which was operating about 300 miles east of Luzon in the Philippine Sea) into the heart of the typhoon. The carriers had been conducting raids against Japanese airfields in the Philippines and ships were being refueled. However, due to worsening weather, attempts to refuel generally failed.

Because of over one hundred mile-per-hour winds, very high seas, and torrential rains, nine warships were severely damaged, three warships capsized and sank, over one hundred aircraft were wrecked or washed overboard, and the aircraft carrier USS *Monterey* was forced to contend with a very serious fire caused by an aircraft striking a bulkhead. Within all of this damage a total of 1,450 lives were lost. Losses to the 3rd Fleet was so severe that Admiral Chester Nimitz commented the typhoon's impact was a crippling blow which might be compared to the suffering in a major military action.

After discovering this information in my research, then looking at the sea maps of 1944, then reading the weather forecasts given to the South Pacific military of the timeframe, and applying some of my own logic, I came to an interesting conclusion. The storm my father encountered was most likely the increasing energy surge building up and heading westward into the Philippine Sea, and would end up becoming Typhoon Cobra.

In December 1944, there were no hurricane or typhoon tracking technologies and no advance storm warning systems as are known and in operation today. The disastrous event of Typhoon Cobra spawned the need for storm tracking and advance storm warning systems to be developed. By mid 1945, the U.S government had installed a workable grid of tracking devices in the Pacific Ocean for tracking storms, and also providing its fleet of ships with updated advance storm warnings.

Even though my father suffered from combat fatigue for many years, it was my understanding from him later in life that he never sought assistance for the problem from the Veterans Administration (VA) because of his mistrust of them. About five years after his discharge from the army, the VA lowered his thirty-percent disability to ten-percent disability, and he never trusted the VA after that action. Until he died at age sixty-eight, his VA disability remained at the ten-percent level. My father was dedicated to the United States of America and its principles; those symbolized by the flag of this great country – that of *the Red, White, and Blue*. He was, however, very much distrustful in many of its politicians and government officials. He always said that in spite of the country's faults and poor politicians, it was still the ***greatest*** place to live on the face of the earth.

Some weeks after my father died, I was going through his belongings he left for me. In one of the boxes were some of his things he brought back from the war. Among those things (which I still have) were a Japanese Rising Sun Flag, some miscellaneous Japanese soldier's field pack items, Chief Clay's tribal necklace, some expended bullets, some island intell photos, Ryan's Bible, Dad's Bible, one of Dad's dog tags, some of Dad's metals, and a bayonet. Interestingly, the inside cover of Ryan's Bible was inscribed with nearly all the same duty stations that my father had been. There was a note stating Ryan's family could not be found. All of my location attempts for Ryan failed as well. My sister has our father's other dog tag and some of his other metals. An interesting note is that our father was only awarded one Battle Star, yet he was in three major battles. Both Dad and Mom always stated this, and I uncovered the proof while writing this book.

After I entered the military and began accumulating awards, I inquired of my father why he never went back and sought the award of the other two Battle Stars. He always shrugged the inquiries off stating the efforts were useless. Then came a day after years of this reaction, I challenged his avoidance. I was home on leave and we were in his garage one afternoon looking at some tools he had bought when my question was reissued. My father stopped his actions and gazed out the garage window for a few moments as if he were lost in another world somewhere. When finally returning, he said, 'Son, metals are just fool's glory. They don't mean nothin'. I was over there and I know what I saw. I ain't diggin' up the past for worthless tin stars, and I ain't talkin' about this anymore.' The faraway look in his eyes ceased to fade for quite a while, and he did not speak for much of the remainder of the afternoon either.

The topic of his lost Battle Stars was never mentioned again, and I respected its avoidance.

Some other interesting fragments from my father's WWII days were uncovered only a few weeks prior to the publishing of this book. While searching for a few items of military memorabilia long stored away, I came across an old cigar box which once belonged to my father. He stored many of his treasured keep-sakes in those old boxes. Looking through items in the box, my eyes met with great surprise to see a 'Voice-O-Graph' from the WWII era. Picking it up, greater surprise met my eyes to find three more of them underneath it! All four of the cardboard records were intact, but had sustained a few cracks over the years. Maybe one day I'll discover a company with the technology for possibly 'light-scribing' the imprints from the groves. Then my father's words from nearly seventy years ago will live again.

As for the wedding rings my father hand carved from bayonets, my sister was given our mother's ring, and I was given my father's ring. As of the writing of this book my sister still has her ring in safe keeping. My ring was worn through many military missions until one day it was placed on the finger of a woman I was to wed. Then, as time passed, the ring was placed in safe keeping. As the years would unfold, a court order would unfairly adjudicate the ring from my ownership. Thus, with that wrongful action, I would lose it and its concrete ties to my part of aesthetic family history forever. It was that legal action and the elements of this story which caused me to arrive at a firm conclusion – there are some courts having no conscience of fairness whatsoever with their passing of judgments.

Yet, in spite of some poor politicians and some poor judges, there is still something tremendously great about this country called the United Stated of America. There is still no other nation like it on the face of this earth, there never has been, and, in my opinion, there will never be until Christ sets up His Kingdom! It is for that fact *alone* that American soldiers still proudly serve in its military and their footprints of *the Red, White, and Blue* are known everywhere on this globe. God bless the USA!

So many warriors over centuries of time have enacted similar actions as my father did when finally landing on the soil of their native homelands. So many warriors over centuries of time have uttered similar cries as my father did when finally reaching their beloved home fires and its families within. Having been a warrior myself, having been among warriors, and even being the son of a warrior, we may hold high and proud the banners of God and country – but loved ones are the ones we are coming back to.

We may fight and die for concepts and precepts, but families are those whom we embrace the most. Their memories help keep us alive in the wastelands of netherworlds. God, country, and family are the prime ingredients for the warrior's Semper Fi.

Infantry fighting on Leyte in late October 1944

Robert Adams and his wife Pauline.
Photo was taken a few weeks after
Robert returned home from WWII in 1945

Home Connections

Most dictionaries define 'home' as a place of residence, or a dwelling place for a person(s) to live in. Further expounded on, home is the main place in which living things come into and depart from. It is the main location from which living things operate their lives. From a human perspective, home should incorporate many nurturing elements for developing the lives of its residents. Home *should* be a place in which its members find safety, security, feel protected, and above all else – feel loved. This entry is not a psychology lesson. However, those previous words were for providing some background setting for this entry.

In all of my military travels, I longed for returning either to my home base, or when having a home, for returning back home. All military members share that same desire. Missions in other places may be exciting and intriguing; however, home is where the heart is, and home is where one remains free. Home has unique connections other places cannot provide, or even compare with.

Consider some elements written within the previous entries of *Finally Made It Home* and *Glory Days Epilogue* as paralleling within the lives of all military members. Soldiers feel patriotic, leave home, train for war (either in defense of their home front, or their allies), engage the enemy in combat, and hope to return home someday.

Compare now those elements of the previously scripted true-life story of my father with another true-life story – that of God, His Son (Christ Jesus), and all of humanity. God loved mankind so much that He gave Christ Jesus in death as a means of sacrifice for restoring us wayward people back to Himself (reference John 3:16). Christ *voluntarily* lowered Himself from His home in Heaven and come to planet Earth. He took on human form and lived for thirty-three years just as everyone else did of His time. He fought with the devil and conducted an innocent ministry for the good of humanity. As a result, He was arrested, sentenced to capital punishment, and placed on death row a very short time before being executed.

Christ Jesus knew what it was like to leave His home of splendor for temporary duty in a land foreign to the things of His world. Christ knew what it was like to engage in warfare for winning the freedom of people in bondage (humanity held in sins). Christ knew what it was like to return back to His home base (Heaven) and His Commanding Officer (God the Father) and His family (the angels). Christ successfully accomplished His mission to Earth; because He loved humanity enough to give His life for us all. Actually, that type of love God which has for us all is *true love* – and it is of the high grade caliber of *Divine red, white, and blue*. *Red* for the blood Christ shed for everyone on the cross; *white* for the innocence Christ was for all of humanity that none of us could never be; and *blue* for the eternal love of God Almighty. God's *red, white, and blue* always equal restoration; and on His Divine level, will always eternally seal the soul.

On the human side of the restoration equation process, we must navigate through the minefields of evil during this journey of life. It is a spiritual warfare. The Bible portrays this many times throughout its pages. Jesus mentioned this struggle many times, of which Matthew 6:24 is one of them. The Apostle Paul even delineated possessing spiritual armor for fighting spiritual battles (reference Ephesians 6:10-18). The struggles we humans go through in this life are most difficult, regardless of whether one is a Christian or not. However, the Christian has a resource of assistance, which the non-Christian does not have – the Lord God.

In going through the struggles and heartaches of the warfare of this life, loved ones pass on to the other side. They, in Christ, with all Heaven has in store, wait for us there. When they cross over to the other side, they have loved ones from years before waiting on them. When our loved ones crossed over, they, like my father returning home from WWII, more than likely fell to their knees in Heaven and, with loved ones gathering all around them, cried something like, 'I finally made it home!' They exchanged worlds; this earthly one for a heavenly one. And, in doing so, they made a much better and greater ***home connection***. They made a *solid contact to a final life association* – an *eternal home with God*!

One day it will come time for us, who are alive and are war weary from combating the strife of this life, to make that transition for crossing over to the other side.

Those of you reading this book, who HAVE God in your life and have your names written in Christ's Book of Life because you accepted Him, your entry will be met with angels, family, and friends. You will have passed from the warfare of this life; its shadows will have molded you into a place of honor. The Lord God said in 1 Samuel 2:30, *"...For those who honor Me I will honor..."* and you will then be ready to be received by Heaven's Host.

Those of you reading this book, who DO NOT HAVE God in your life and do not have your names written in Christ's Book of Life, please, as we say in the military, 'let's do this' and 'get it done.' That one decision for Christ Jesus will make all the eternal difference you could ever imagine. Then one day you, too, will say, 'I finally made it home!' *"You therefore must endure hardship as a good soldier of Jesus Christ. No one engaged in warfare entangles himself with the affairs of this life, that he may please Him who enlisted him as a soldier"* (II Timothy 2:3-4).

In Harm's Way

In 1968, Mel was a U.S. Department of Defense Navy Civilian, working as a WD-10 Aircraft Electrician. He was assigned at a Naval Air Rework Facility which refurbished the U.S. Navy and Marine Corps A-4 Skyhawk fighter aircraft. Navy depot level rework facilities accomplish depot overhaul for all of their weapon systems. At the time, the Marine Corps needed temporary depot level support of their A-4 aircraft in the field at a particular Marine Aircraft Group (MAG) unit in Vietnam.

At the depot facility, a small group of civilian volunteers were assembled. After briefings and field supplies were issued, the group was flown to the MAG airbase southeast of Da Nang for a deployment lasting six months. Military civilian personnel from the depot detachment were dressed in sanitized military uniforms and worked alongside the Marine aircraft technicians. Their mission was to provide the Marine Corps with limited depot level maintenance support in the battle field. While at the MAG, one of the supply items issued the military civilians was an M-16 rifle. The Marines felt it was the duty of the military civilians to defend the airbase, and themselves, in the event of an enemy attack. Hopefully, though, this would never happen while the military civilians were assigned there; but, it did.

About four months into Mel's deployment was when it happened.

Mel and a couple of Marines were securing from working very late one night on an aircraft, when shots were heard at the edge of the airfield. Their aircraft was the farthest out and, therefore closest to the incoming assault. All three men jumped behind a bunker wall as one Marine immediately radioed for back-up, while the other Marine and Mel returned gunfire. Before back-up arrived, the enemy had breached the inside airfield perimeter and was closing in on the two Marines and Mel. Luckily, when the back-up Marines did arrive, their presence repelled the assault.

In 1991, Arthur was a U.S. Department of Defense Navy Civilian, working as a WD-12 Aircraft Sheet Metal Mechanic. He was assigned at a Naval Air Rework Facility which refurbished U.S. Navy and Marine Corps aircraft and components. In support of Desert Storm, the U.S. Navy and Marine Corps needed temporary depot level support of their H-53 aircraft in the field in Saudi Arabia.

At various depot facilities, groups of civilian volunteers were assembled. After briefings and field supplies were issued, the groups were flown to field repair points in Saudi Arabia for varying lengths of deployment. Military civilian personnel from the depot detachment were dressed in sanitized military uniforms and worked right along with the U.S. Navy and Marine aircraft technicians. Their mission was to provide the military with limited depot level maintenance support in battlefield conditions. While at the field repair point, one of the supply items issued the military civilians were various items of weaponry. The military deemed it was the duty of the military civilians to defend their repair bases, and themselves, in the event of an enemy attack. Hopefully, though, this would never happen while the military civilians were assigned there, but; it did.

About six weeks into Arthur's deployment was when it happened.

One afternoon Arthur was assisting some Marines with testing for cracks in an airframe. From out of nowhere came a small band of insurgents led by a few Iraqi soldiers charging their position. Arthur and the Marines jumped for cover of the bunker while the Marines quickly responded with their M-16s and he began firing with his pistol. Back-up soon arrived, some of the enemy were killed, and the remaining others fled.

In 2002, Greg was a U.S. Department of Defense Army Civilian, working as a GS-13 Weapons Acquisition Specialist. He was assigned to the U.S. Army Materiel Systems Command in Washington, D.C. In support of the War on Terror, all branches of the military needed immediate, temporary, and long-term, logistics support of their weapons and supply systems in the field.

At various bases and facilities, groups of civilian volunteers were gathered and assembled. After briefings and field supplies were issued, the groups were flown to various assigned field staffing units in the world. Greg was sent to Saudi Arabia for a six month deployment, but ended up staying for eight months. Military civilian personnel in Greg's detachment were dressed in sanitized military uniforms and worked right along with the U.S. Army soldiers. Their mission was to provide the military with detailed logistic support in the battle field. While in field, one of the supply items issued the military civilians in Greg's unit were various items of weaponry. The military deemed it was the duty of the military civilians to defend their base, and themselves, in the event of an enemy attack. Hopefully, though, this would never happen while the military civilians were assigned there. There were a few times, though, events became a little 'difficult' and Greg was close to drawing his pistol, but he never had to.

These three brief stories of military civilians putting themselves in harm's way for the USA are but a few. There are thousands more on record. Patriotism for *the red, white, and blue* of the USA runs very deep. It runs far beyond its leaders; it runs very deep into the fabric of its founding framework and the soul of its peoples. Patriotism for *the red, white, and blue* of the USA goes far beyond the military uniform; its runs mysteriously deep for the love of God, family, country – and the will to be free.

At some point during my military career, I served with all three of the men in this story. Mel and Arthur had served in the military on active duty. Greg was never on active duty because of medical reasons. Yet, he wanted to serve his country so much that he became an army civilian. Greg is retired from the army, and the fate of Mel and Arthur is unknown. All three of these men had told me they would serve the USA until retiring unless death took them out first. In my opinion, that is loyal devotion to *the red, white, and blue;* yet, most of their patriotic dedication was accomplished in an unusually thought of status – while as *military civilians*.

As the technological base of America's weapons systems rapidly increases, so also must increased technical support be required from the depots and logistics commands. In the future, more and more military civilians will be seen donning sanitized military uniforms, and defending *the red, white, and blue* – while placing themselves in harm's way.

God Bless them, the military,
and the USA!

No God...

Charles Martel was born on 23 August 686 AD in Hertstal, Belgium (formerly known as Heristal). Martel served as a Frankish military and political leader, served as Mayor of the Palace under the Merovingian Kings, and ruled *de facto* during an interregnum reign at the end of his life using the title, Duke and Prince of the Franks (737 to 743 AD). In 739 AD he was offered the title of Consul by the Pope of Rome. Martel, though, refused the position.

Charles Martel was a most intriguingly significant figure for his people and in the development of European history. Yet, he will always be most famous for one great battle, The Battle of Tours. Its victory earned him the name of 'The Hammer' or 'Charles the Hammer' from then, and on into the pages of history books.

The religion of Christianity had begun blazing through Europe after the crucifixion and resurrection of Jesus Christ. In the back draft of its evangelistic flames, another religion arose. That religion was Islam and was catapulted by the fervor for war its clerics had acquired from its aspiring new leader – Muhammed ibn Abdullah (which is sometimes transliterated as Mohammed). The aim of Islam under Mohammed was (and still is) submit or die, and their battle cries became *kill the infidels*. This was (and still is) such a different evangelistic concept from the religion of Christianity proclaimed by Jesus Christ – one of peace, unity, forgiveness, love, and restoration.

In 732 AD, Abd-er-Rahman, the Islamic Governor of Spain, led an immense army crossed, the Pyrenees, overcame the forces of Duke Eudes, and advanced as far as the Loire; while raiding, looting, and burning everything as he traveled. All of this pillaging was for one main purpose, the advancement of Islam by force – submit or die. History has always proven that with the 'evangelic' advancement of Islam, comes mass socialistic enslavement. By the way, the Islamic clerics conveniently fail in omitting those 'minor' details. With radical Islam, it is simply convert or die.

In October of 732, Charles Martel met and engaged Abd-er-Rahman for battle outside of Tours. There Martel defeated and slew him in a battle (the Battle of Tours, Poitiers; which is also called the Battle of Poitiers). This battle remains one of the greatest events in the history of the world, for upon its decisive issue depended whether Christianity should continue, or if Islam would prevail throughout Europe. It was this battle, as was previously stated, that gave Charles Martel his nickname 'The Hammer' or 'Charles the Hammer' because of the merciless way in which he smote the enemy.

Now arising again, in this 21st Century is the threat of Islam dominance of the world; and this time by radical Islamic terrorist activity. The world witnessed streaming television news casts of Islamic terrorists impacting planes into the World Trade Center Towers of New York City and the U.S. Pentagon on 11 September 2001. Those horrific events propelled the USA and many other nations into a War on Terrorism. That war, in its initial attack on 11 September 2001, was supposedly instigated by the radical Islamic terrorist leader Osama Bin Laden. He was supposedly found and killed in May of 2011; his death inciting rage across the Islamic world. As of the writing and publishing of this book, the War on Terrorism is still being waged worldwide.

At the end of this entry is an interesting photo. It is the back side of an American one-dollar bill. There is where can be found the phrase *In God We Trust*, and someone has stamped another phrase – *No God But Allah*. This phrase could have been stamped by an impartial individual desiring to incite trouble. However, with the War on Terrorism occurring globally, this is highly doubtful. In light of current troubling terrorist activities taking place within the USA and around the world, highly suspect is the radical Islamic community for the stamping of USA currency.

The radical Islamic community, fueled by the Society of Muslim Brothers, or Muslim Brotherhood (MB), has rekindled its zeal for worldwide dominance by Islam. Radical Islam, fueled by the MB, does not desire that others choose their religious path – they insist the entire world be subject to its oppression.

The MB has a credo of 'God [Allah] is our objective; the Quran is our constitution, the Prophet (Mohammed) is our leader; Jihad is our way; and death for the sake of God [Allah] is the highest of our aspirations.' The MB is bent toward forcing their views of radical Islamic beliefs onto every society on Earth. Two of those basic concepts are: (1) Armed Jihad, in which their movement makes war on everyone else for subduing them; and (2) Shariah Law, in which men can have multiple wives and women have no voice whatsoever in society.

The MB has openly declared their goal of eliminating the country of Israel and annihilating all Jewish and Christian peoples. This mission of 'salvation' is quite diabolically different (and opposed) to the one preached by Jesus Christ. Radical Islam mandates kill the infidels if they do not convert (Surah 9:5 of the Quran), whereas Jesus Christ said He would not cast anyone out (reference John 6:37). Jesus Christ stated in John 3:16, *'For God so loved the world that He gave His only begotten Son, that whoever believes in Him should not perish but have everlasting life.'* Those words proclaim a salvation for everyone with freedom, who desires it, and that freedom comes with the liberty to make that choice. Those words do not come compelled with any threats of cursing and death as does with the religion of Islam.

In order to provide a comparison of Scriptural views, the following very few passages are from the Quran (Koran), and the Holy Bible:

- On unbelievers is the curse of Allah (Surah 2:161)
- Muslims are harsh against the unbelievers, merciful to one another (Surah 48:25).
- Fighting is obligatory for you, much as you dislike it (Surah 2:216).
- Fight against such of those who have been given the Scripture as believe not in Allah nor the Last Day, and forbid not that which Allah hath forbidden by His messenger, and follow not the Religion of Truth, until they pay the Jizya with willing submission, and feel themselves subdued (Surah 9:20).
- ...fight and slay the pagans wherever ye find them, and seize them, beleaguer them, and lie in wait for them in every stratagem... (Surah 9:5).
- *For the mountains shall depart and the hills be removed, but My kindness shall not depart from you, nor shall My covenant of peace be removed, says the LORD, who has mercy on you* (Isaiah 54:10).
- *Through the LORD's mercies we are not consumed, because His compassions fail not. They are new every morning; great is Your faithfulness* (Lamentations 3:22-23).
- *Blessed are the peacemakers, for they shall be called the sons of God* (Matthew 5:9).
- *...what man is there among you who, if his son asks for bread, will give him a stone? Or if he asks a fish, will he give him a serpent? If you then, being evil, know how to give good gifts to your children, how much more will your Father who is in heaven give good things to those who ask Him!* (Matthew 7:9-11).
- *...the one who comes to Me I will by no means cast out* (John 6:37).

Mohammed was born 26 April 570 AD, orphaned at any early age, and then cared for by an uncle. He worked as a sheppard and a merchant and spent much of his time alone in meditation to his god, called Allah.

During his hours of meditation, Mohammed supposedly received the revelations which have become the Quran. Mohammed claimed that Allah ordered him to fight unbelievers until they worshipped Allah. Mohammed said, 'I have been ordered to fight with the people till they say, none has the right to be worshipped but Allah' (Al Bukhari 4:196). An interesting note for adding at this point is that many individuals regard Mohammed as a founder of the Muslim religion of Islam. Wise Muslims know what careful research will uncover; ironclad proof that the religion of Islam existed long before Mohammed was ever born.

Mohammed just knew how to market and launch its concepts very well – after he *added concepts to its doctrine*.

Any military or political strategist is well aware of what any charismatic ***visionary*** can bring into societies – ***purpose***. Purpose, under a great captivating visionary, can meld disconnected and diffracted tribes of peoples into a solid cohesive band of brothers. The preaching of religion can be one of many catalysts for the melting pot of the visionary. It is my opinion Mohammed was very aware of this marketing ploy. History records he was a merchant and, as such, he employed combining that skill with religion for uniting his diffracted tribal kinsman. Then Mohammed used another strategy for cohesion of the masses; deflect its aggression. He saw the evangelistic prolific spread of Christianity in his area of the world as a threat. Therefore, he turned the collective afflictions and aggression within the diffracted tribes of his kinsman toward Christianity as a common enemy.

Mohammed achieved success as a new leader, both religious and otherwise.

Nazi Leader and German Führer Adolf Hitler accomplished basically that same strategy during the late 1930s and during WWII. Hitler deflected some of the woes of Germany at the Jews, became very successful, and rose to power rapidly. Johnny-come-lately visionaries who come preaching a new religion, coupled with aiming hatred and aggression toward another party (or peoples) is a very deceptive combination – and can be dangerously costly in a variety of ways.

The devil is very well aware of this tactic. He's had centuries of experience with it – on this globe and in the heavens as well.

As previously stated, with radical Islam, the religion believes submit or die. Their vision is global Jihad. Additionally, if any Muslim converts to Christianity, those traitors are also targeted for being killed. Apparently, it is their way or no way, and they have no problem sending infidels straight to hell without any remorse...wow, where's the love in the radical religion of Islam? Such a cultural-socialistic concept has the very same fiber from which dictators, despots, tyrants, and autocrats are made from.

On the flip side from the concept with Mohammed as a sheppard, merchant, and religious leader, is another *Sheppard* – Jesus Christ. Christ was born supposedly on 25 December about 2000 years ago. His actual birth date is unknown, but closest estimates have been somewhere between 5 and 8 January. For various reasons (which have conflicted through the ages), His birth date of 25 December was set by Pope Julius in 350 AD. Christ's life became so significant in changing world history that its events marked the beginning of a new calendar, and thus the annotation of Anno Domini (AD) was instituted. The term Anno Domini is Medieval Latin, and is commonly translated as 'In the year of the Lord.' However, it is more specifically correct as Anno Domini Nostri Iesu Christi (In the Year of Our Lord Jesus Christ).

Jesus Christ was the Divine Son of God having been born of the Virgin Mary, was raised by His earthly parents and never orphaned. Christ was carpenter by vocation, and True Sheppard in ministry to the needs of people. The last three years of His life, He was in ministry proclaiming the love of God and healing the sick. *Christ never advocated taking up armed warfare against any one*. This was, and still is, quite the opposite spectrum from Mohammed's distorted idealistic concepts of fighting until the infidels submit or die.

One leader had in his mind the unification for the hordes of diffracted tribes. Another Leader had the love of God for lost souls, and the restoration and for of all mankind embedded within His heart. One leader ***sought change*** with swinging swords and submit or die. Another Leader ***brought transformation*** through peaceful teachings, healings, and forgiveness. One leader died a natural death and is still in his crypt. Another Leader also died, and by a horrendous execution; but rose back to life again, and disappeared into the heavens stating He would return someday. One leader is revered as a founder of religion, which actually existed long before him. The other Leader is generally mocked by the entire world for founding a peaceful faith, which is actually the only concept that can set people free from *any and every* bondage.

Why, readers may ask, am I writing these things?

It is simply this:

American GIs have sacrificed themselves on the altars of the world for freedom from the tyranny of its despots for numerous decades.

Terrorists, and *especially* the radical Islamic community, are apparently bent on bringing their conflict *within* the once-secure borders of the USA homeland.

It is high time for American citizens to get involved in voting out useless politicians.

It is high time for American citizens to get involved in local and national government issues more than ever before.

It is high time for **ALL** Americans to insist on equality for everyone – not just those entities deemed difficult to understand.

As for me, the author of this book, I am becoming very concerned if there will come a time in the future similar as the Battle of Tours on American soil.

And if so, how soon, and for whom will the sanction of those actions be cast?

God Bless the USA!

If left up to Allah
(and if there be such)
he would subvert it!

Taps

The song *Taps* is always played at military facilities at bedtime, usually about 2200 (10 PM). *Taps* signifies the day's ending; the finality of all things accomplished during that day. It further signifies the coming for a time of rest at the ending of activities, so that more activities can commence at morning's dawn, or when '*Reveille*' is sounded.

Taps is also always played at military funerals. If anyone has ever attended a military funeral or has viewed one, solemn respect can be seen rendered to the fallen comrade when *Taps* is played. Long after the ceremonies have ended and faded, we can still hear the haunting melodious tunes rising from the song of *Taps* resounding within our minds. It is the solemn lonely sounds of that song which will usually bring tears to our eyes.

Most individuals, though, have never heard the true story behind the humble beginnings of that song.

Reportedly, the song's roots began in 1862 during the American Civil War, when Union Army Captain Robert Elli was with his men near Harrison's Landing in Virginia. The Confederate Army was on the opposite side of the narrow strip of land. During the night, Captain Elli heard the moans of a soldier who lay severely wounded on the battlefield. Not knowing if the moaning was emitted from a Union or Confederate soldier, the captain decided on risking his life and bringing the wounded man back for medical attention. Crawling on his stomach through gunfire, the captain reached the wounded soldier and began pulling him back toward his encampment.

When the captain finally reached his own lines, he discovered the man was actually a Confederate soldier, but the soldier was dead. The captain lit a lantern, suddenly caught his breath, and shocking numbness set in. In the dim light, a familiar soldier's face glared back at him – it was that of his own son. His son had been studying music in the Southern States when the war began. Without telling his father, the son had enlisted in the Confederate Army.

The following morning the heartbroken father asked permission of his superiors to give his son a full military burial, despite his enemy status. His request was considered and only partially granted. The captain had asked if he could have a group of army band members play a funeral dirge for his son at the funeral. His request was denied since the soldier was a Confederate. But, out of respect for the father, they did permit him the use of just one musician. The captain chose a bugler and he requested the bugler to play a series of musical notes he had found on a piece of paper in the pocket of his son's uniform.

The wish was granted; and now that special haunting melody is commonly known as the song *Taps,* which is played at military funerals. Its words are as follows:

Day is done, gone the sun, from the lakes, from the hills, from the sky, all is well.

Safely rest, God is nigh, fading light, dims the sight. And a star, gems the sky, gleaming bright, from afar, drawing nigh, falls the night.

Thanks and praise, for our days, neath the sun, neath the stars, neath the sky, as we go, this we know, God is nigh.

General Douglas MacArthur once said, 'Old soldiers never die, they just fade away.' That famous quote was spoken by him in his farewell address to Congress when he was being relieved of command after WWII. His comment was actually a fitting way in saying a form of 'Taps' to a highly successful and much decorated career while one is still living.

On a comparable note as the end of this book draws toward a close, what will be the 'Taps' for your life, reader of this book? Will you fade away from the life of this world, and reignite into God's great Heaven? Or, will you just dim from sight and forever dwell within the shadows of evil's hellacious dark nights?

God does have a remedy for the shadows of evil's eternal dark night nightmares!

All that is needed is for His Son to light the way. Once we accept God's Son into our lives, all inner stains of wrongs are erased forever, and a charter for a new life will begin. Taps will be sounded for the former person fading away inside. Then, all of Heaven rejoices over the new person having just come into the Light of God.

When '*Taps*' are sounded, if the 'SonShine' has ignited the inside of that life, it will reignite into a burning eternal flame.

Surrender

In 1975 the U.S. Army Unit that I was assigned to was called in for patrolling and guarding an area devastated by Hurricane Eloise. The Fort Walton Beach to Panama City area of Florida was a disastrous mess for several weeks. No electrical power, buildings blown down (and away), roads and streets buckled and caved in; the region was crippled for a long time with cars, trucks, and boats strewn and piled up everywhere.

Two other soldiers and I were dispatched on duty one night to guard a shopping center. One of our unit trucks dropped us off at the curfew hours of 2100, and we were to be there all night until 0600 the next morning. The shopping center was a horseshoe shaped strip-mall which had sustained heavy damage. Several cars and yachts were piled up and strewn over its parking lot.

Around 0100 in the morning, we three soldiers were in the middle of the parking lot near a couple of wrecked cars. Sitting on the pavement with our backs leaned against one another, we had just finished a 'delicious' can of C-rations. One of the soldiers stood to his feet and, as he picked up his M-16, ducked quickly down onto the pavement. As he came back down, the sound of an engine was heard as dim light was seen cast onto the wrecked cars near us. We three soldiers then scurried for cover next to the nearest car.

Peering through the broken glass of the overturned car, we saw a large black or dark blue car with two occupants entering the shopping center. The car stopped at the first place of business, the passenger exited and walked to the store. He then appeared to be checking on something at the storefront. Then, the car would drive to the next establishment as the one walking around would check something at each storefront and write something on a pad. This same routine was repeated many times. With those two civilians being out over four hours past curfew, and their car tag being from another state, something was clearly not quite right with all this. We three came to a huddle for a quick action plan, and then, in the cover of the darkness, stealthy dispersed in three different directions.

Near the end of the shopping center loop, the man doing the walking was turning from a darkened area of a business store. Instantaneously, we three soldiers descended from the darkness on those two men; with our M-16s aimed at them, we shouted, 'Halt!' One soldier dropped his weapon on the man in the car (from the passenger side), and the other soldier and myself dropped our weapons on the walking man. I was in position for covering any false move from either perpetrator.

The walking man immediately stretched both arms and hands straight up and high as he could, while at the same time dropping the pen and pad and screaming several times, 'Don't kill me!' The car driver immediately, but slowly, just extended both arms and hands outward from his sides as far as he could reach them. Both men had the palms of their hands fully open.

Once we had them arrested, and checked their IDs, we then detained them while waiting for the Sergeant of the Guard to make his rounds at 0200. The two men had an excuse of being business owners and checking on the damages to their properties. However, they were violating curfew and, by imposed military law, we had to arrest them. When the Sergeant of the Guard arrived, he radioed for police back-up, and both men were taken into custody. Later information was provided to us, that background discovery revealed those two men were part of an organized theft ring from another city. Our arrest of those two men led to the police authorities of different states arresting further members of their theft organization.

What is the universal sign or symbol for surrender? It is very simply both hands up! And both of them empty, too! One does not play around with those situations unless they have a death wish, intoxicated with some substance, or are just plain stupid. Reasonable people do not challenge guns pointed in their faces.

In combatant operations, when faced with surrendering situations, would-be captives do not hold arms and hands partially high (unless wounded) – but as high as possible. The would-be captives do not clasp anything in their hands; the palms of their hands are as open as possible. Why? Their captors must observe their sincerity for surrender! If arms and hands are not held as high as possible, there may arise opportunities for the availability of reaching hidden weapons which could be used in last moments of dying offense. If palms are not fully open, they may hide a weapon which could also be used in last moments of dying offense. False surrender, or lack of full surrender, can take the captive to an early grave.

In another story of military surrender, all hostile actions of World War Two (WWII) were officially and formally ceased with the Surrender of Japanese forces on the deck of Battleship USS *Missouri* on 2 September 1945. The 'Mighty Mo' weighed 58,000 tons fully loaded; had a crew of 2,700 men; had three main sixteen inch guns (three sixty-seven foot barrels per turret); could fire 2,700 pound rounds twenty-three miles in fifty seconds with pin-point accuracy; had armor plating varying from 13.5 to 17 inches thick; and was five feet longer and eighteen feet wider than the RMS *Titanic*. The 'Mighty Mo' was known as America's most formidable ever battleship.

At the time of the signing for the surrender of WWII in the Pacific, the captain for USS *Missouri* was Captain Stuart Murray. On the USS *Missouri's* website, is documented some memoirs by Admiral Stuart Murray. Admiral Murray stated that General MacArthur ordered the least time possible for the surrender ceremony to be accomplished; within thirty minutes if possible. For about a week, though, before the surrender signing on 2 September, the USS *Missouri* took on Japanese emissaries.

The Japanese were permitted to have a news reel photographer, and he was assigned to a 40-mm gun platform. In consideration of a possibility for a kamikaze hero with the cameraman, two Marines were assigned to the photographer. Each Marine had one hand on the photographer's leg, and their other hand on the butt of their holstered Colt 45. The photographer was shivering and shaking so much that the pictures were thought to be probably not much good. There was no question, though; the cameraman was kept well under control.

On the deck of the USS *Missouri* is a commemorative metal plate inscribed 'Over this spot on 2 September 1945, the instrument of formal surrender of Japan to the Allied Powers was signed, thus bringing to a close the Second World War.' The USS *Missouri* remained docked in Tokyo Bay for three days following the Japanese surrender of their forces of WWII. Thus, its presence displayed an illustration of power; a show of force after surrender.

Although the Japanese Empire signed the Unconditional Surrender terms on 2 September 1945, aboard Battleship Missouri, the historic beginning of the end came on board the Destroyer USS *Levy* about a week earlier on 22 August 1945. That ship was host to the first formal surrender of Japanese occupied territories in the Mid Pacific Ocean as WWII was ending. The geographic location was Wake Island, in a lagoon of the little-known, but strategically important Mille Atoll, in the Marshall Islands group of the Mid Pacific Ocean. Thus, the Wake Island surrender was actually the beginning of the formal surrender of WWII in the Pacific.

Shift from the topic of surrender in military combatant roles, to that of Christianity. What does the act of surrender involve for the Christian? Interestingly, there are some Biblical principles which apply with surrender, and yes, the military combatant principles previously mentioned do have some somber spiritual applications as well.

God is not the captor who may send captives to early graves – it is our own foolish actions in the devil's war of worlds. Additionally, it is our own will that refuses surrender to God which *prohibits* His best workings in our lives. Quite simply, our pride negates His masterful plans.

In considering the symbol for universal surrender, think about the song, *Rock of Ages*. One of the lines in that song is *'Nothing in my hands I bring, simply to the cross I cling!'* Surrender does not hold onto things of the past, but will let go of them. Surrender realizes the futility of wasted efforts. Surrender realizes continued efforts can create a vast wasteland of nothing. Surrender relinquishes one's rights to something of value.

One of the best concepts about the topic of surrender to the Lord God is contained with the song, *I Surrender All*. One of the lines in that song is *'I surrender all, I surrender all, all to Thee my blessed Savior, I surrender all.'* The song speaks of things to give up, and lay at His feet, for being totally consumed with Christ. The author of that song obtained it from his prayers of yielding his life to God in whatever God would have him to do in service for Him. Many years later, after writing the song, he greatly influenced the lives of many seminary students in his college classes. One of those students became the world-wide known Evangelist, Reverend Billy Graham.

The price of personal ***surrender*** may appear great for the present, and its lasting effects change only one restless soul – however, its impact can go onward for altering the world of tomorrow.

In a different perspective of surrender, there is a story in Chapter 19 of Matthew about a Rich Young Ruler who came to Jesus. His presence with the Master was for wanting to know what more could he do for guaranteeing his eternal life. The story is also found in Chapter 10 of Mark, and Chapter 18 of Luke. Those other passages contain a little more information for adding details to the overall story, because the events were seen from differing perspectives of its writers.

According to the Bible, the young ruler apparently possessed much wealth and was obedient to the Law of Moses. He had come running to the roadside where Jesus was for gaining an audience with Him in front of the crowd of people; a trait shared by many politicians and public figures. In the dialogue with Christ, all he willing offered was something of his *own works* as assurance for eternal security. He did not need anything else for he basically had it all.

That young ruler had the freedom his office gave him to accomplish whatever he needed, and whenever he desired. This immense authority combined with his riches, equated to great power; an influence that became a dominance within him, which subdued his very soul. That rich young ruler became *spiritually incapacitated* because of the *supremacy* his possessions gained within his life; they had become his god. He could not let them go. He could not ***surrender*** them to the one true God of everything, and he was aware of it.

His running to Christ at the roadside that day was actually a glorified plea-bargain attempt in a spiritual sense. His plea-bargain was in hopes of maintaining some possessiveness within the affluence which controlled him. He wanted to bargain with The God, for keeping at least one little puppet god in the closet of his life.

Christ Jesus, being of the God Trinity, and knowing all things, saw right into the motives lying hidden within the ruler's heartland. Jesus laid the groundwork of the religious Law with the Commandments, and then went straight to the core material of the man's heart to perform surgery – but the patient denied access.

That rich young ruler was actually a patient-in-waiting carrying spiritual bondage. He had come to the Great Physician, but offering his own cures. Then, when faced with the correct diagnosis and cure, he refused Divinely Professional Treatment. He came to Jesus in a 'boxed-in' life situation, and departed the Master's presence even more boxed-in. In walking away from Christ, he sadly discovered his riches had taken greater hold on him with such a grip, that he could not let go of them.

By not letting go of what Christ asked of him, he was announcing to everyone that those things were his god, because they meant more to him than the real one true God. By not letting go of those things, he made a political announcement to his people – that his material status meant more to him than the people did, and even God Himself. His love of wealth actually cost him popularity in the eyes of his people, which was most likely not realized when he first walked away.

Lack of the ability to ***surrender*** cost this man much more than he ever thought it would. Non-surrender to the Lord God cost this man esteem in the eyes of his people. Most likely, he was never respected as much ever again, paralleling his declining popularity ratings. His wealth most likely dwindled to some degree. If he had some form of business, when people heard of how he refused giving to them, and the Master, they moved their business dealings elsewhere. The list of impacts can go on and on, but the fact remains – lack of surrender to God, *will* cost something in the end.

Let's take a look at Biblical principles within two Biblical characters; the first one is Job. He lost his family and everything at no fault of his own. He was an innocent victim of the devil. One day Job literally found himself sitting in ashes scraping his sores with broken pottery. Throughout all his troubles and trials, he came to place his total confidence in God with saying, *'Though He* [God] *slay me yet will I trust Him'* (Job 13:15). That trust Job was in reference to, was actually a surrendering confidence. Surrender, in some situations, is when all is lost, yet total confidence (faith) remains fixed (stands firm) in God for a better day in the future. Surrender can generate from becoming overwhelmed with great love of restoration power. Surrender can come from the ***ashes*** of being made humble.

The second example in a Biblical character is Jonah. God called him to go and minister of His grace and salvation to the Assyrians. Jonah absolutely refused. No, they don't deserve redemption, blah, blah, blah, and on and on would be Jonah's excuses for having things his own way. Jonah ran from his calling and, in his rebellious exit, caused dismay to the ship and crew he had boarded. Collateral damage can sometimes occur to innocent bystanders because of our rebellion. Jonah realized this and sacrificed himself for being thrown overboard. But, God had a seaworthy rescue agent in the form of a great fish.

There are many who believe Jonah and the great fish is a fairy tale. Think again.

Berosus, a fourth century B.C. Babylonian historian, recorded many traditions of the Assyrians where Jonah was directed by God to go minister. One of those Assyrian pagan traditions was the worship of a fish-man, who came up out of the sea. During the time of Jonah, the people of Nineveh believed a divinity, part fish and part man, would send them messages after rising out of the sea.

Oh, you can be assured that when Jonah 'arrived in style' after vomited up on the beach from a great fish, they undoubtedly were very receptive to anything Jonah had to say. God's timing and ways are perfect, are they not? From Jonah's willing mouth and their willing ears, it was actually a case of *double surrender*! Then when the missionary job was done, Jonah went off by himself to whine for a while. Sometimes total surrender is intermittent and takes incremental stages to complete.

Surrender can be generated from becoming overwhelmed with great deliverance and restoration power. Surrender can come from *being made humble after searching and going through* pits of darkness. Surrender can come from being left alone and being asked to come back home in humility. Sincere desire creates the *base material* for surrender.

Military organizations undergoing a process known as Strategic Redirection (SR) are given new missions, and sometimes new names. Many times they are unrecognizable from their former mission because their overall *function* has been completely altered. In a personal life, an SR can have the same effect; and God can impact it spiritually also. Engaging God, by total surrender to Him, can result in a life with a new mission, a new purpose; and sometimes it is not recognizable from its former state.

Just as the surrender of WWII in the Pacific actually began a week earlier on 22 August 1945, on Wake Island, personal surrender always begins at some point before the final action of total surrender. Surrender is usually always incremental; its *compliant path* is seen coming long before the *submissive destiny* arrives.

Recall the USS *Missouri* docking in Tokyo Bay for three days after the surrender of WWII? Also, recall the metal plate on the deck of the USS *Missouri*, commemorating the surrender of WWII? Part of the reason was for an illustration of power, a show of subdued force, an intimidation factor. In life, we all do much the same in various ways to our adversaries. We show them the power we have over them in court, in policies, in relationship interactions, et cetera. We will hold dates, events, words, and so forth, as markers for proving to them we have supremacy. We want them to know they have been put in their place beneath us – and by us.

Consequently, God does not display intimidation factors in order to subdue us (unless we are headed toward destruction). God chooses love for winning us over to Him. Love can change more decisions than any method of controls, any laws, or any tyrannical powers. Love places no markers in a life that a certain spot is where they defeated another. Victories or successes are always shared by love. Chapter 13 of I Corinthians expounds the many ways of love's nature.

Pride will keep commemorating the places of subdued victory, and say, 'This is where I overcame you. Never forget it.' Love will take the other back to that special place and say, 'This is where we first met. I will always remember it.' Love will often reminisce of that place and time, and the surrounding circumstances. Interestingly, most Christians can say, 'This is the time and place where I first came to know the Lord.'

What happens when one surrenders ***ALL*** of themselves to the Lord God? One's relationship with God deepens and becomes more personal than ever before. One's relationship with their friends and family deepens. One's connectivity with their fellowman improves (relationships; jobs; associations). One's life in general seems enriched; because one has increased inner peace, increased confidence in God (faith), increased success rate of events, decreased stress, improved health (usually lower blood pressure), and life becomes easier and a joy worth living. Christians who never smile have never surrendered. They probably harbor some hidden resentment against God.

Total surrender for the Christian places them on the pathway to becoming one with God. Total surrender changes everything and is the basic platform from which all spiritual operations can successfully be carried out. Surrender is not 'all about me.' Surrender is about humility. Surrender will consider others before oneself. Surrender is a strategic redirection in one's life – I'm done; I'm finished; I can't go on this way anymore.

Surrender to the Lord God creates freedom like has never before been experienced – ever.

"But now, O LORD, You are our Father; we are the clay, and You our Potter; and all we are the work of Your hand" (Isaiah 64:8). *"What shall I render to the LORD for all His benefits toward me?"* (Psalm 116:12).

The author during his army days.

What If?

With many of the stories in this book, I have given credit to the Lord God. The reason is very simple: I believe in God, and I give Him credit where credit is due. With my own personal stories in this book, I have given credit to the Lord God for delivering me. Some may say those things were just chance with the randomness of the factors of life. Then if so, how does one explain the dream of my mother the night before my head injury? How does one explain the vision coming to my friend while driving the morning of my head injury? How does one explain those skull faces and the grim reaper coming to me just a few months before my traumatic head injury?

The mystic would not discount those strange events as mere chance – and neither should the agnostic or the Christian. Within those events are some elements of truth, for those dreams and visions foretold a later reality. For many years I have been a Christian; I've not been perfect, but as a Christian I have been forgiven of my imperfections. All through my years as a Christian I have seen God prove Himself to me that He is real. The stories in this book are but a few of His proofs given for me. My other books contain many more stories of why I believe in the Lord God.

For those now reading this book, who may not believe in God, I put forth this question to you – What if it's true? What if there really is a God?

That 'truth' opens up a whole realm of reality both for this world and the next; and thus, interweaving the physical and spiritual dimensions. If there is a God, then is that God interested in us – humanity? If there is, and if He is, then to what extent would that God go toward assisting us along our difficult pathways in life?

There are many answers to that question, and this book addresses two of those. One is briefly discussing a fundamental concept of faith in God and is in this chapter. Another has been revealed with three of my strange life encounters; that God is definitely interested in our sojourning of this road we call life. He is interested enough to inform us, and others, of what may happen to us. He is also compassionate enough to provide rescue for us when there is no hope at all. And now, for the faith factor, examine a few interesting concepts.

When finding one's way (especially when lost in any situation) it is very important to know *who* to believe, *what* to believe in, and *when* to make the right decisions. When traveling and becoming lost, an incorrect map will not be of any benefit whatsoever. Having the map of a location other than the area where we are located is totally useless. Likewise, what about a belief in God? Some believe in God, and some do not. Some believe that because we cannot see God, He does not exist. Some believe that a good God would not let bad things happen at all.

The wind blows, and we do not see it; but we *feel* it and we *know* it is there. Our eyes cannot 'see' the realm for the energy wavelengths of the winds. Yet, our skin can detect the pressure of those wavelengths of energy. One of our senses cannot detect the presence of the winds, *yet* another one of the senses empathically detects the *evidence* of its presence. Such is the same with God. We may not be enabled to see Him; however, the evidence of His existence is present, and has been around for eons of time. The question really is then, 'Do we have the *faith to believe* the evidence He has presented?'

This is a fallen world, the dominion of satan and, therefore, evil things *will* happen – ***<u>even</u>*** to good people. Such does not negate the existence of God, but actually proves it. All through the events with the existence of a fallen world are scattered occurrences of God's intervention for extraction, improvement, redirection, comfort, forgiveness, etc. God has chosen to prove Himself in the evil corridors of a fallen world. It is in those avenues of darkness that His Light shines on the pathways for life's traveler to navigate correctly and efficiently.

It is up to each individual life traveler to access the Light given them, to determine if there is a God, and what He means to them personally. It is with that Light whereby the life traveler sojourns a safe navigation home to Heaven after this life. Without embracing that Light, the life traveler will grope in darkness and never find God.

There is constant rhetoric about whether the Holy Bible is truth or just fables. Is it a work of fiction or non-fiction? This stuff has been around in various forms since the Garden of Eden, and it will not cease until God brings an end to the age of man. It is appalling to discover the high numbers of people in this world who believe the Bible is just a bunch of hand-me-down stories, only good for soothing the soul at bedtime, and not actually correcting its wrongs.

Jewish candidates of antiquity entering the Rabbi ranks were required to memorize the entire Torah, the first five books of the Bible. The societies of today's world fail in comprehending that many Jewish Rabbis know all of the Talmud's 2,700 plus pages perfectly by heart; word for word. The Talmud is a compilation of legal opinions and debates and also contains the first five books of the Bible. Many of today's Jewish Rabbis are very unimpressed by the computerization of the

Talmud, for it means very little to have it in any computer memory drive, when it can be embedded within their own living memory banks. This memorization task is word for word, and is taught and administered by their Rabbis. It is totally illogical for even considering any errors in the Torah when every Rabbi in the nation had (and still has) the Torah completely and flawlessly memorized.

As centuries passed, more tales of errors in the hand-me-down stories of the Bible became commonplace. Then came the Dead Sea Scrolls, and their discovery cast away many shadows of Biblical errors. The Dead Sea Scrolls proved the accuracy of the transmission of the Bible. Those scrolls were discovered in Qumran Cave One near the Dead Sea in 1947. There are in existence some Old Testament manuscripts dating back to about 150 BC. Comparing those manuscripts with the Dead Sea Scrolls proved to be basically the same, with only minor variations due to penmanship slips and spelling variants. The fact that those manuscripts were distanced by nearly two thousand years and were basically the same is an indicator of the fantastic accuracy for the Bible's manuscript-copying transcription efforts from antiquity.

An entire copy of the Book of Isaiah was also discovered at Qumran. Even though the copy of Isaiah discovered at Qumran Cave was a thousand years earlier than the oldest dated manuscript previously known (980 AD), it proved ninety-five percent word-for-word identical with the standard Hebrew Bible text. The five percent variation was, again, mainly pen slips and variations in spelling.

Manuscript discoveries like the Dead Sea Scrolls provide undeniable evidence that today's Bible Scriptures are (for practical purposes) exactly the same as when originally inspired from God and recorded by His Prophets. The Dead Sea Scrolls prove that the scribes (copyists) of Biblical manuscripts took great care in going about their work. Those ancient scribes knew they were duplicating God's Word. Therefore, they treated it with great respect and went to all extents possible in preventing error from entering their literary works. Those scribes carefully counted every line, word, syllable, and letter for complete accuracy. They realized that someday they would give an account at God's Judgment for their penmanship.

As for just the soothing, feel-good, hand-me-down, stories of the Bible; consider just two of those stories. One is Jonah in the belly of the fish, and the other is Noah with the big ole boat. Most of the world scoffs at these stories being so ridiculously untrue; better rethink them again. Jesus referred to both of those men as truthful events when He spoke in Matthew 12:40 and 24:37. Therefore, if those events were not factually true, why would Jesus reference them as true? There are stories of Jonah in other cultures. His story is also written in the Koran (Qu'ran), and the Bahá'í faith (Bahá'í is a monotheistic religion founded by Bahá'u'lláh in nineteenth-century Persia). Therefore, in that whale of a story, there must actually be something to it for other ancient cultures having recorded it. The most common story of Jonah is written in the Jewish Bible, from which Christianity obtained it.

Jonah, though, is not the only person who has been purportedly swallowed by a great fish, and lived. There are also legends about men being swallowed by whales. One story is about James Barley, who in 1891, while sailing aboard the whaling ship *Star of The East*; reportedly fell overboard; was supposedly swallowed alive by a whale; was recovered later; and lived. Another story is of a Marshall Jenkins, who in 1771, while sailing aboard the whaling ship *Edgartown*, came into contact with a whale. The whale broke one of the smaller whaling boats in two, took Jenkins into its mouth, and submerged with him. Later returning to the surface, the whale vomited Jenkins onto the wreckage of the broken boat. Jenkins was very bruised but not seriously injured. Those big fish stories cannot be substantiated as factually truthful. Why? One major fault is that depending on where one obtains research information, much of the resulting data is very conflicting. Some data even states the ship's logs do not even list those sailors as having served with them.

There are two very interesting facts for consideration, though, about the Jonah Story. One comes from Berosus, a 4th century BC Babylonian historian, and the other appears much later in history. Berosus recorded many traditions of the Assyrians where Jonah was directed by God to go minister. One of those Assyrian traditions was the worship of a fish-man, who came up out of the sea. During the time of Jonah, the people of Nineveh believed in a divinity sending them messages by a person rising out of the sea. They believed he would be part fish and part man. Oh, you can be assured that when Jonah 'arrived in style' being vomited up on the beach from a great fish, they undoubtedly were very receptive to anything Jonah had to say. God's timing and ways are perfect, are they not?

The other interesting fact is the preservation and discovery of the name 'Yunas' in the ruins of ancient Nineveh. In 1847 the British actually began limited excavation of Mosul, or ancient Nineveh, and intermittently continued the task for decades. Nineveh had seven entry gates within its wall; one name was known for centuries as 'Neby Yunas.' When modern archeologists unearthed a mound at that gate, beneath it was found the ruined palaces of Nineveh's kings. Also was found the tomb of Neby Yunus, which means 'Prophet Jonah' (Yunus and Yunas are variations in the spelling of Jonah), thus, confirming the historicity of the Jonah Story.

Whichever way God transported Jonah to Nineveh, it for sure had something to do with a great fish. If Jonah were swallowed by a great fish (which I believe), then

God could have temporarily shut down the gastric system of the fish (which has for centuries given skeptics problems with this story). God could have also extended the available oxygen Jonah needed for his stay in that fish. Who knows, God could have even lowered Jonah's metabolic system similar to a hibernation-like state.

As for that ridiculous big ole boat that Noah could not possibly build (and 'load all them' animals on it), there is an Internet website which shows some interesting photos of the remains of Noah's Ark. Go to the specialtyinterests.net website and search for the remains of Noah's Ark on Mount Ararat in Turkey. The resulting data and photographs are very surprising. May I strongly suggest readers take a look, and read over the recorded documentaries? Whichever way God chose in Noah building that Ark, God made sure it was big enough and strong enough for carrying ***all*** the animals God wanted on it ***at the time***. Consider all the differing races and types of people that came from just those eight people of Noah's family. That same concept can also be applied to the animals as well.

Consider that Jesus came from God, and God is truth, and God cannot lie. Therefore, Jesus cannot lie. As a result, the stories of Jonah and Noah are, thus, validated as true by the words of Jesus, the Son of God, aka Mr. Universe of Truth. The Dead Sea Scrolls were actually God Proofs of His existence, just waiting for modern day archeologists to discover them for the skeptics. The finds and photographs of Jonah's Tomb, the Gate in Nineveh, and Noah's Ark in Turkey are God's imprints for the skeptics over time. It has been said that seeing is believing, yet with some things, some individuals still cannot believe, even with a print from God.

In further pondering the question of 'What If?' there is a fact which needs serious consideration. An interesting fact for increasing faith that there is a God is not found in the Bible, but actually within mathematical science. Most everyone in this age has heard the phrase 'Do the math.' It implies the application of sound rationale to a problem and arriving at a logical conclusion. Following that train of thought, there is an actual mathematical probability model of Christ Jesus being who He claimed Himself to be. Christ Jesus claimed to be the long prophesied Jewish Messiah, the Son of the living God.

Dr. Pete Stoner was a Christian Mathematical Scientist in the 1950s and 1960s. He was the Department Chairman of both Mathematics and Astronomy at Pasadena City College until 1953; Chairman of the Science Division, Westmont College, from 1953-57; Professor Emeritus of Science, Westmont College; and Professor Emeritus of Mathematics and Astronomy, Pasadena City College. Dr. Stoner wanted to look at the statistical probability of Jesus being who He said He was.

Dr. Stoner once took the elements of sixty prophesies within over three hundred references contained in the Old Testament declaring the coming of, character of, and the life and death events of the Lord Jesus Christ, and programmed them into his computer. In the magazine *Science Speaks* during the mid 1960s, Dr. Stoner revealed those resultant findings. He wrote that using the laws of mathematical probabilities, there exists a 10^{157} possibility that someone could fulfill all of those prophesies in an average lifespan of about seventy years. The algebraic number 10^{157} is 1 with 157 zeros following after it, and would look like the following number:

10,000,000,000,000,000,000,000,000,000,000,000,
000,000,000,000,000,000,000,000,000,000,000,000,
000,000,000,000,000,000,000,000,000,000,000,000,
000,000,000,000,000,000,000,000,000,000,000,000,
000,000,000,000,000

There has never been that number of people born on the face of the earth –ever– and will never be for eons of time to come. Yet Jesus was born about 2000 years ago and fulfilled all the requirements of those prophecies as spoken about Him in the Holy Bible. Dr. Stoner projected that if only eight prophecies were fulfilled, the statistical probability would be 10^{17}. Using a silver dollar for each of the 10^{17} numbers, silver dollars could cover the State of Texas two feet deep. Then if one of those silver dollars contained a special marking and one person was given just one chance to find the one specially marked silver dollar – that would be the one chance in 10^{17}.

A 10^{157} probability with silver dollars would be much the same, only with those two-foot deep silver dollars extending to cover the entire Unites States. The website for this information is found within the geocites.com website. Conduct a search on Science Speaks and Doctor Peter Stoner. These facts are also documented in the book *More Than A Carpenter* by Josh McDowell.

Jesus, however, accomplished all of those prophecies in His life span of just thirty-three years. Therefore, the statistical probability was actually increased to more than ***double*** that 10^{157} possibility that someone could fulfill all of those prophesies! So then, 'do the math' for yourself. Look at the statistical chances of Christ being who He said He was: the centuries-long prophesied Messiah. The facts of statistical math point to Christ Jesus being the Son of the one true living God. However, just doing the math and accepting the stats are not the guaranteed tickets for entry through Heaven's Pearly Gates. Asking forgiveness for all our wrongs and sins, and accepting Christ into our heart and life as Savior and Lord, and believing – **IS** the guaranteed ticket.

There is definitely a factual ingredient here with the math of 10^{157} for building faith in the fact of there being a God. God is a mathematical genius. He ran the numbers on His Son Christ for modern-day mankind to see. Yet, it is up to modern man for acceptance of those stats into their hearts. Then, just as the unseen wind can be felt – the unseen things of *factual statistics* can be revealed through the eyes of faith, for making the things of God a reality!

There is another interesting fact worth examining. This fact is found in the statistics of literary science. Ever thought about how the Holy Bible ranks in manuscript comparison with other writings of antiquity? In the field of literary science, part of the validation process for ancient manuscripts (MSS), is by the numbers of originally hand-scribed manuscripts identical to the earliest known original handwritten manuscript. Those manuscripts are not mimeographed, photographed, or photocopied in any form, but actual ancient hand-scribed copies.

The book of *Caesar* has ten copies from antiquity spanning about 1,000 years; it has an MSS of 10. Plato's *Tetralogies* has seven copies from antiquity spanning about 1,200 years; it has an MSS of 7. Tacitus' *Annals* has an MSS of 20 spanning 1,000 years. Pliny The Younger's *History* has an MSS of 7 spanning about 750 years. Suetonius' *DeVita Caesarum* has an MSS of 8 spanning about 800 years. Homer's *Iliad* has an MSS of 643 spanning about 500 years.

The MSS number and span of years vary with the various sources. In my illustration I wish to focus on Homer's *Iliad* for the literary world seems to hold that work has the most validated authentic piece of writing from antiquity. For Homer's *Iliad* I'll use an MSS of 700, because in my readings, research results have revealed an MSS ranging from about 640 to just over 800. Therefore, an MSS of 700 would be a good average number for this illustration.

Continuing on, there are approximately 700 manuscripts of Homer's *Iliad,* which have been hand-scribed in antiquity, and are basically identical to the earliest known copy. Homer's *Iliad* has been widely accepted in the literary world as the oldest most validated authentic writing from antiquity. Thus, no one doubts the authenticity of Homer's original work.

Apply that very same literary scientific principle to the Old Testament of the Bible, and the results will come up to an MSS of approximately 8,000, over 1,600 years. Extend that concept to the New Testament and the results reveal an MSS of about 25,000, **AND**, its span of time only ranges about 50 years! Quite a significant difference!

Logic from these statistics alone would dictate that the Holy Bible is, in fact, the most validated authentic writing from antiquity that we have in the world. However, most people do not want to accept these statistical facts. Why? Consider the ole cliché, which says, 'I know what was said (or I know what I want), don't confuse me with the facts.' We just simply want things our own way, which does not include God's way most of the time.

An MSS of 700 spanning about 500 years, versus an MSS of 8,000 spanning 1,600 years, and an MSS of 25,000 spanning 50 years – is quite a significant validation factor for authentication! The MSS concept for the Bible is a very solid faith-building ingredient. It very factually illustrates the soundness for the uniquely high authenticity of the Holy Bible, **AND**, having faith in what it says! Sadly, though, too many people do not accept those sound facts. They prefer living lives of 'not being confused by the facts.'

In mentioning the preference of living 'not being confused by the facts,' why is it that God has to be proven, but everything else can be accepted seemingly Carte Blanc at face value? Why does the one true God, creator of the entire universe, have to prove His Divinity? Yet, all the other pagan gods of this world are accepted as is. Why does faith in God require empirical truths, proofs, and reasons? Yet, having faith in any other thing appears so easy, and even acceptable? Why does it seem that the God of all gods, must show, and prove, some Certification of Divinity? Yet, all the other lesser gods receive lavish credentials of supernatural origin, and most of those gods are just figments of imagination or carved icons.

It seems from the very beginning of mankind's existence, man has held high court on proving the existence of God. God has been on trial since mankind rebelled against Him. Just go out on a clear starry night, and look up into the heavens. Where did all those stars come from? Some entity, far more advanced than any of us, flung all of it into place long before this sphere of Earth came into existence. That entity is God. Quite simply, gazing into a telescope or a microscope for a while at the countless wonders those lenses can reveal, should settle the matter of God's existence.

This is the intersection where faith connects with reality; we need His guidance and we need to trust in Him. God always knows what is best, for He sees the future and we cannot. In order to arrive at the best destinations of life by excellently-lighted pathways, there must be something to make the connections. That something is called faith. Faith is invisible as the wind, it but touches the heart of God.

One last consideration on whether or not there is a God – the ***dedicated minister***. When one becomes a dedicated minister for God, not one who plays religion or uses it only as a vocation, that one will definitely find out the existence of evil. Satan, and all those on his side are forever assailing God's dedicated ministers. If there was a doubt in the mind about there being a God, becoming one of God's ***dedicated*** ministers will definitely erase that doubt.

If there is a devil for the evil,
then there must be a God for the good.

"The fool has said in his heart,
'There is no God…'"
(Psalm 14:1)

Epilogue

With this epilogue, take a brief memory glance into some past entries within this book, which hopefully will add gravity to its ending. In the shadows of the stories explored within these pages, consider the term Semper Fidelus, which is Latin for 'Always Faithful.' It is the well-known motto of the U.S. Marine Corps, and the phrase is often shortened to just Semper Fi. In maritime contexts, this motto dates back at least to the fourteenth century. It has served as a slogan for many families and entities, in many countries. However, for the U.S. Marine Corps they have embraced a unique view of Semper Fi as their motto. It doesn't mean 'Sometimes Faithful' or 'Usually Faithful' – it means 'Always Faithful.' It is not negotiable; it is not relative; it is absolute.

Christ Jesus, the Son of God, is just that – Semper Fi, Always Faithful. One reason is because He said, *"...and lo* [notice, take note, listen], *I am with you always, even to the end of the age"* (Matthew 28:20). The Lord God also spoke through the Psalmist, *"Forever, O LORD, Your word is settled in heaven. Your faithfulness endures to all generations..."* (Psalm 119:89-90). Friends and relatives may turn away and desert us. Programs, benefits, entitlements, societies, governments, and kingdoms may falter and collapse, ultimately leaving us deserted and alone. But, Christ Jesus will never leave us alone. He is always faithful; He is *Semper Fidelus*, or *Semper Fi*.

In considering the attribute of faithfulness, contemplate and meditate on the Bible Scriptures found in the addendum of this book. When seriously considering those Bible Scriptures, quite simply there is only ONE conclusion; the Lord God is *Semper Fi – He is Always Faithful*. Once co-joined into the heart of anyone accepting Him, He sets up residence and will never leave.

Howie Day recorded the most popular version of the song *You and I Collide.* The song relates that even though two people are different, they come crashing together at times. Similarly, we crash into God and ricochet ourselves away. In our careening departures, the Lord God still remains Semper Fi, even throughout all of our crashing-into-glancing-away times with Him. All through this life, we always collide with the Lord God, and sometimes actually do connect with Him for a time. For those who believe in spiritual dreams, those times of colliding connections are especially prominent in dreams. In dreams of the night we are still and cannot resist God. It is then when He can freely speak to His people.

About 2000 years ago, God incarnate came to this Earth in the person of Christ Jesus. It was then a trilogy of powers collided; God, satan, and mankind. It was then God demonstrated superiority over evil, and the coming eternal defeat of satan. It was then God demonstrated the depth of His profound love for mankind. Ponder the fact that Christ gave His life in crucifixion for all of mankind, and why He would do such. Then consider after accomplishing what He did, would He become as the gods the pagans and deists believe in – leave us, forget us, and not display any faithfulness to his creation? No, the Lord God Almighty would NOT, because His *true colors ran blood red* over 2000 years ago. If there is any parting of ways, we are the ones who depart; yet the Lord God remains *Semper Fi – Always Faithful.*

Consider some stories from this book. The life of my grandfather was spared by soldiers digging him from a pile of death. The life of my father was spared by Chief Clay risking his own and shooting a sniper. My father risked his own life many times to save the lives of wounded soldiers. George was spared with his buddy's heroic actions; Perkins selflessly gave his life for George. George received honors sixty-five years late, but never once complained. Roy became a casualty of war, and never once complained. Finley saw the aftermath of war while working reconstruction efforts; and its impact carved a life of ministry. Chad was knocked down and bleeding, but was not taken completely out; his wounds caused him to yield toward God's calling of becoming a preacher.

Those are only a few sparse examples. History books and diaries are chocked full of events, facts, and testimonies lending to patriotic acts toward one's country, fellow soldiers, and especially in the altruistic sense of dying for the beliefs of God, family, and country. In times of war, there are tragic losses on every front. Over periods of time toxicity could rise in the ranks, and within the staffing offices for the battlefields, over ALL that has transpired. Undue blame can spread like wildfire because perceived victories may have been delayed or lost.

It is during those times that soldiers, who remain fixed and engaged for their cause, are fighting at their finest hours. Some will even opt for becoming 'the fall guy' and give of themselves that others may live, or allow others to receive the greater recognition. Those are the ones who see a greater victory ahead just over the horizon. Those are the soldiers who have gone from meritorious defense episodes, toward seizing the day with a Semper Fi disposition. During conflict's darkest hours, they seize the times with *the best choices*, and will secure a brighter dawning tomorrow for everyone.

In parallel, Christ Jesus is, in general, somewhat similar to soldier hero images. Through Christ's crucifixion, He took the blame for every one of our transgressions. In His most trying time of earthly life, Christ became 'the fall guy' for every horrendous thing we have ever done. Through Christ's sufferings, at His and the world's darkest hour, He seized the day and times with a Semper Fi disposition, securing a brighter dawning future for anyone who will accept His payment for them. Anyone, choosing to ask for forgiveness and accepting Christ's payment for them, will seize the day and thus secure a brighter emerging eternity.

Rewind and recall some of the stories in this book: *Crushed, Lost at Sea, Glory Days, Reconstruction Duty, Helmet Hit*, and maybe some others. Why would God do such for average GIs serving *the red, white, and blue*? God loves with a perfect love we cannot understand. His love is limitless as stated in Psalm 57:10, *"For Your mercy* [love] *reaches unto the heavens..."* His love extends out from Him, and to everyone equally the same as stated in John 3:16, *"For God so loved the world that He gave His only begotten Son, that whoever believes in Him, should not perish but have everlasting life".* God, therefore, is an equal opportunity God. Anyone still not believing that John 3:16 parallels a Semper Fi God, just recall the two stories of *Crushed* and *Lost at Sea*. I was the undeserved recipient of miracles sent from God in those two events for sure!

In relating many of the stories in this book (and some with spiritual applications), over many years, I have seen many individuals in tears return home to God. Some of those were even involved with the occult. Others have found hope after lengthy periods of desperation. The stories in this book, with their spiritual applications, are significant examples of God's love and faithfulness to people. They impart the message of how God has fallen in love with us (very deeply in love), and we are His prize creation.

The Lord God Almighty
is a Semper Fi God

Postscript

Postscripts are usually afterthoughts written to any scripted work. This postscript centers on the concept with postage stamp sized images contained in the book cover and within some pages of this book. When gazing at any collection of military imprints, those images express a *specialized message*; one which indicates those guardians have been protectorates for their home front. There may be numerous other messages and intentions expressed, but that *one specialized message* is the focal point of them all.

Beyond many of the primary intentions, those specialized images further go on toward expressing a concept seldom thought of. Those images were proof the sentinels had actually been on guard, either on the home front or in those distant outposts of foreign lands. Those images were proof of those sentries having left their foot prints, boot prints, blood, sweat, and tears on the soils of those places. Those images were proof that the patriots of the *red, white, and blue* had been there acting as shields against tyranny for the oppressed gripped in bondage.

Look again at another side of life's cultural coins with this postscript analogy…

Every one of us human beings are traveling through this life and leaving personalized tracks of where each of us *has been*. We leave traceable markings everywhere we go. At times, we leave indelible imprints on the lives of others. They can be anywhere from scratches to scars, or from approval to adoration. Whatever the striking impressions, we leave imprints. Some imprints last lifetimes; some for the good, and some for the bad. Regardless of what we do in life, we are going to give an account of those actions and activities some day in the distant future. That day is called Judgment Day and is mentioned many times in the Holy Bible, specifically in Revelation 20:11-15. On that day, Heaven's Books will be opened and each of us will be judged according to what is recorded in them. Each and every one of us writes a book with our thoughts and actions as we go through this life. Heaven records those things in the books which are kept on each of us.

Jesus said, *"For there is nothing hidden which will not be revealed, nor has anything been kept secret but that it should come to light"* (Mark 4:22). None of us know the scripting styles of Heaven's recording methods, but those recordings will be accomplished flawlessly. Those details of our 'recorded works' will reflect the inner nature of who we actually were in this life. Those details will be snapshots of our 'life works' reflecting the *indelible footprints of the soul.* Those imprints represent what we *actually were* in this life. If the life record is tarnished, that record can be cleaned up with Christ Jesus, the Son of God, before leaving this life. He transforms the sinner into a saint – a Christian. Christ Jesus is the best cleaner of sinners and Christians of all time; He is the 'C-Cleaner' of all the centuries.

"For the wages of sin is death, but the gift of God is eternal life in Christ Jesus our Lord...For whoever calls on the name of the Lord shall be saved" (Romans 6:23 and Romans 10:13).

Addendum One
Foxhole Faith

Finding God and accepting Him into your life in a legal sense is actually somewhat of a Pre-trial Hearing. Each one of us will stand before God some day and give an account of the things we have done in this life (reference Revelation 20:11-15). By asking God for forgiveness and accepting it before that time through His Son, Christ Jesus, the record book Heaven keeps on each of us is expunged of our transgressions. As we continue living in the Grace of God, ask forgiveness, and receive it, our record is continually expunged. Thus, we in effect, live out somewhat of a Pre-Trial Hearing until Judgment Day. Then, Christ, as our Attorney, will stand in for us.

Finding God and accepting His Son, Christ Jesus, is not complicated at all; it is actually very easy. It is as simple as '**ABC**.' The first of the ABCs is ***Admit***. Admit having sinned, being sorry for them, and being in need of God's forgiveness for those attitudes and actions not meeting God's standards. *"If we say that we have no sin, we deceive ourselves, and the truth is not in us. If we confess our sins, He is faithful and just to forgive us our sins and to cleanse us from all unrighteousness"* (1 John 1:8 and 9). *"For the wages of sin is death, but the gift of God is eternal life in Christ Jesus our Lord"* (Romans 6:23).

The second of the ABCs is ***Believe***. Believe that Christ Jesus died and rose again paying the penalty of death for our sins. God loves us so much, He provided a way of escaping that penalty. *"For God so loved the world that He gave His only begotten Son, that whoever believes in Him should not perish but have everlasting life"* (John 3:16). *"But as many as received Him, to them He gave the right to become children of God, to those who believe in His name"* (John 1:12).

The third of the ABCs is ***Confess***. Confess Christ Jesus as the Lord and Savior of your life. Confessing Jesus means yielding total control of our lives to Christ and telling someone else what He has done. If He is given ownership of our life, He becomes our new manager. *"...if you confess with your mouth the Lord Jesus and believe in your heart that God has raised Him from the dead, you will be saved...For whoever calls on the name of the Lord shall be saved"* (Romans 10:9 and 13).

A simple prayer for finding God can be as follows: 'Oh God, I have sinned and am sorry for them, and I am in need of You. Please forgive me. I believe Your Son, Christ Jesus, died and rose again for my sins. I confess Him now as Lord and Savior of my life. Thank you for saving me.' This simple prayer does not 'save' a person in and of itself. What is does, though, is provide the basic foundation for expressing one's faith to God in saving one from their transgressions, through the sacrifice Christ gave of Himself with His own death on the cross about 2000 years ago.

After saying that prayer, do two things preferably every day. First, read some in the Holy Bible, and second, pray. Praying is just simply talking to God and is communication with Him. Prayer begins a relationship with God and nourishes it. Bible reading is further nourishment for the spiritual person reborn inside from the former person. It lays a solid foundation of knowledge and wisdom for building the personal relationship with God.

Then, find a good Bible teaching church to attend. Christians need other Christians around them, praying for one another, and gathering themselves together worshipping God. Regular church attendance further nourishes the new life begun with God. Churches are like people in some respects – they have personalities. Find a church where your life can blend in and where the Spirit of God is freely flowing.

Addendum Two
Foxhole Passages

A few passages from the Holy Bible For guidance through life's pathways, and while in the foxhole...

1. *...Be strong and of good courage; do not be afraid, nor be dismayed, for the LORD your God will be with you wherever you go* (Joshua 1:9).
2. *But You, O LORD, are a shield for me, my glory and the One who lifts up my head (Psalm 3:3).*
3. *I will both lie down in peace, and sleep; for You alone, O LORD, make me dwell in safety* (Psalm 4:8).
4. *...He does not forget the cry of the humble* (Psalm 9:12).
5. *Yea, though I walk through the valley of the shadow of death, I will fear no evil; for You are with me; Your rod and Your staff, they comfort me* (Psalm 23:4).
6. *Though an army may encamp against me, my heart shall not fear; thought war may rise against me, in this I will be confident* (Psalm 27:3).
7. *As far as the east is from the west, so far has He removed our transgressions from us. As a father pities his children, so the LORD pities them that fear Him. For He knows our frame; He remembers that we are dust* (Psalm 103:12, 13 and 14).
8. *Your word is a lamp to my feet and a light to my path* (Psalm 119:105).
9. *How precious also are Your thoughts to me, O God! How great is the sum of them! If I should count them, they would be more in number than the sand; when I awake, I am still with You* (Psalm 139:17 & 18).
10. *All we like sheep have gone astray; we have turned, every one, to his own way; and the LORD has laid on Him the iniquity of us all* (Isaiah 53:6).
11. *Through the LORD's mercies we are not consumed, because His compassions fail not. They are new*

every morning: great is Your faithfulness (Lamentations 3:22 & 23).

12. *For I have no pleasure in the death of one who dies,' says the* LORD GOD. *'Therefore turn and live!'* (Ezekiel 18:32).
13. *...what man is there among you who, if his son asks for bread, will he give him a stone? Or if he asks for a fish, will he give him a serpent? If you then, being evil* [of carnal nature], *know how to give good gifts to your children, how much more will your Father who is in heaven give good things to those who ask Him!* (Matthew 7:9-11).
14. *Are not two sparrows sold for a copper coin? And not one of them falls to the ground apart from your Father's will...Do not fear therefore; you are of more value than many sparrows* (Matthew 10:29 & 31).
15. Matthew 11:28 – *Come to Me, all you who labor and are heavy laden, and I will give you rest.*
16. *For God so loved the world that He gave His only begotten Son, that whoever believes in Him should not perish but have everlasting life* (John 3:16).
17. *...the one who comes to Me I will by no means cast out* (John 6:37).
18. *...I have come that they may have life, and that they may have it more abundantly* (John 10:10).
19. *For all have sinned and fall short of the glory of God* (Romans 3:23).
20. *But God demonstrates His own love toward us, in that while we were still sinners, Christ died for us* (Romans 5:8).
21. *If we confess our sins, He is faithful and just to forgive us our sins and to cleanse us from all unrighteousness* (I John 1:9).
22. *And if anyone sins, we have an Advocate with the Father, Jesus Christ the righteous* (I John 2:1).
23. *And God will wipe away every tear from their eyes...* (Revelation 21:4).

Addendum Three
It Is The Veteran

It was the veteran,
not a reporter,
who has given us freedom of the press.

It was the veteran,
not a poet,
who has given us freedom of speech.

It was the veteran,
not a lawyer,
who has given us the right to fair trial.

It was the veteran,
not the campus organizer,
who has given us the right to demonstrate.

And it is the veteran,
who salutes the flag,
who served under the flag,
and whose coffin is draped by the flag!

~ Author Unknown ~

Addendum Four
Notable Military Clichés

When in Doubt, Empty The Magazine

Happiness Is A Belt-Fed Weapon

Sniper Motto:
You Can Run, But You'll Just Die Tired!

Machine Gunner Motto:
Accuracy By Volume

Except For Ending Slavery, Fascism, Nazism, And
Communism, WAR Has Never Solved anything

U.S. Air Force:
Travel Agents To Allah

The Marine Corps:
When It Absolutely, Positively,
Has To Be Destroyed Overnight

Death Comes To Everyone
Marines Can Make The Appointments

What Do I Feel When I Kill A Terrorist?
A Little Recoil

The Marine Corps:
Providing Enemies of America
An Opportunity To Die
For their Country Since 1775

Foot Prints of Red, White, and Blue

Life, Liberty, And The Pursuit
of Anyone Who Threatens It

Spectre Gunners
You Can Run...But You Can't Hide

Special Ops
Anywhere...Anytime

It's God's Job to Forgive Our Enemies
It's Our Job To Arrange The Meetings

One Long Shot
Multiple Kills
U.S. Naval Gun Fire Support

My Kid Fought In Iraq
So Your Kid Can Party In College

A Dead Enemy Is A Peaceful Enemy
Blessed Are The Peacemakers (Matt 5:9)

If You Can Read This, Thank A Teacher
If You Can Read It In English, Thank A Veteran

It Does Not Rain In The Infantry
It Rains On It!

If You Cannot Stand Behind The troops
Please Feel Free To Stand In Front Of Them

Terrorists Are Like King Salmon
Life Is Good Until The Seals Show Up

Some People Spend An Entire Lifetime
Wondering If They Made A Difference In The World
The Military **Does Not** Have That Problem

"For what profit is it to a man
if he gains the whole world,
and loses his own soul?

Or what will a man give in
exchange for his soul?"

~ Matthew 16:26 ~

This book was produced and
published in association with:

Dream Weaver Ministries, LLC
Springfield, Ohio, USA
www.dreamweaverministries.com

Made in the USA
Monee, IL
04 November 2026

67374729R00154